CAMBRIDGE LIBRARY COLLECTION

Books of enduring scholarly value

Physical Sciences

From ancient times, humans have tried to understand the workings of
the world around them. The roots of modern physical science go back to
the very earliest mechanical devices such as levers and rollers, the mixing
of paints and dyes, and the importance of the heavenly bodies in early
religious observance and navigation. The physical sciences as we know them
today began to emerge as independent academic subjects during the early
modern period, in the work of Newton and other 'natural philosophers',
and numerous sub-disciplines developed during the centuries that followed.
This part of the Cambridge Library Collection is devoted to landmark
publications in this area which will be of interest to historians of science
concerned with individual scientists, particular discoveries, and advances in
scientific method, or with the establishment and development of scientific
institutions around the world.

The Recent Progress of Astronomy

Elias Loomis (1811–89), Professor of Mathematics and Natural Philosophy at
the University of the City of New York, published the third edition of this key
work in 1856, at a time when the discipline of astronomy was making rapid
advances. Recent technological progress had led to a phenomenal number
of astronomical discoveries: the existence of a new planet, Neptune; a new
satellite and ring for Saturn; irregularities in the movement of many planets
and stars; thirty-six new asteroids; numerous comets; extensive catalogues of
stars; and new and important observations on the sun. Loomis's report is a
treasure-trove of information regarding these discoveries and the significance
they had at the time. The chapters on the history of American observatories,
various astronomical expeditions, public astronomical surveys, and telescope
manufacturing in the USA provide access to information not otherwise
available. *Recent Progress* is a key text in the history of astronomy.

Cambridge University Press has long been a pioneer in the reissuing of out-of-print titles from its own backlist, producing digital reprints of books that are still sought after by scholars and students but could not be reprinted economically using traditional technology. The Cambridge Library Collection extends this activity to a wider range of books which are still of importance to researchers and professionals, either for the source material they contain, or as landmarks in the history of their academic discipline.

Drawing from the world-renowned collections in the Cambridge University Library, and guided by the advice of experts in each subject area, Cambridge University Press is using state-of-the-art scanning machines in its own Printing House to capture the content of each book selected for inclusion. The files are processed to give a consistently clear, crisp image, and the books finished to the high quality standard for which the Press is recognised around the world. The latest print-on-demand technology ensures that the books will remain available indefinitely, and that orders for single or multiple copies can quickly be supplied.

The Cambridge Library Collection will bring back to life books of enduring scholarly value (including out-of-copyright works originally issued by other publishers) across a wide range of disciplines in the humanities and social sciences and in science and technology.

The Recent Progress of Astronomy

Especially in the United States

Elias Loomis

CAMBRIDGE UNIVERSITY PRESS

Cambridge, New York, Melbourne, Madrid, Cape Town, Singapore,
São Paolo, Delhi, Dubai, Tokyo

Published in the United States of America by Cambridge University Press, New York

www.cambridge.org
Information on this title: www.cambridge.org/9781108013932

© in this compilation Cambridge University Press 2010

This edition first published 1856
This digitally printed version 2010

ISBN 978-1-108-01393-2 Paperback

THE

RECENT PROGRESS

OF

A S T R O N O M Y ;

ESPECIALLY

IN THE UNITED STATES.

BY ELIAS LOOMIS, LL. D.,

PROFESSOR OF MATHEMATICS AND NATURAL PHILOSOPHY IN THE UNIVERSITY OF THE
CITY OF NEW YORK, AND AUTHOR OF A COURSE OF MATHEMATICS.

THIRD EDITION,
MOSTLY RE-WRITTEN, AND MUCH ENLARGED.

NEW YORK:
HARPER & BROTHERS,
329 TO 335 PEARL STREET.

1856.

PREFACE.

THE progress of astronomical discovery was never more rapid than during the last fifteen years. Within this period, the number of known members of the planetary system has been more than doubled. A planet of vast dimensions has been added to our system; thirty-six new asteroids have been discovered; four new satellites have been detected; and a new ring has been added to Saturn.

It is especially gratifying to note the progress which the last few years have witnessed in the United States, both in the facilities for observation, and in the number of active observers. It is but twenty-five years since the first telescope, exceeding those of a portable size, was imported into the United States; and the introduction of meridional instruments of the larger class is of still more recent date. Now we have one telescope which acknowledges no superior; and we have several which would be esteemed worthy of a place in the finest observatories of Europe. We have also numerous meridional instruments, of dimensions adequate to be employed in original research. Our own artists have entered successfully upon the manufacture of refracting telescopes of the largest size, and have received the highest commendation from some of the best judges in Europe. These instruments have not remained wholly unemployed. At the observatories of Washington and Cambridge, extensive catalogues of stars are now in progress; while nearly every known member of our solar system has been repeatedly and carefully observed. These observations are all permanently recorded by a simple touch of the finger upon a key which closes an electric circuit—a method recently introduced at Greenwich observatory, and known everywhere throughout Europe by the distinctive name of the American method.

Numerous discoveries have been the result of this astronomical activity. A large number of comets have been independently discovered on this side of the Atlantic; and among these, three at least were observed here before they were seen in Europe. The two nebulæ which have been longest and best known, were never adequately figured until they were observed by the Messrs. Bond; and the planet Saturn, which for many years was made the subject of special study by the elder Herschel, with his wonderful means of observation, first revealed to the Messrs. Bond a new ring and an eighth satellite. The novel spectacle of a comet divided into two nearly equal portions, was first witnessed in America; and an American observer has added one to the long list of planetary discoveries.

Let us indulge the hope that the future progress of astronomy is to be no less brilliant than the past; and that henceforth America may be even more distinguished for her contributions to science, than for her progress in material power and wealth.

CONTENTS.

CHAPTER I.

RECENT ADDITIONS TO OUR KNOWLEDGE OF THE PLANETARY SYSTEM.

SECTION I.

SECTION II.

SECTION III.

SECTION IV.

SECTION V.

CHAPTER II.

RECENT ADDITIONS TO OUR KNOWLEDGE OF COMETS.

SECTION I.

SECTION II.

SECTION III.

SECTION IV.

SECTION V.

SECTION VI.

SECTION VII.

CHAPTER III.

ADDITIONS TO OUR KNOWLEDGE OF FIXED STARS AND NEBULÆ.

SECTION I.

SECTION II.

SECTION III.

SECTION IV.

SECTION V.

SECTION VI.

CHAPTER I.

RECENT ADDITIONS TO OUR KNOWLEDGE OF THE
PLANETARY SYSTEM.

SECTION I.

THE DISCOVERY OF THE PLANET NEPTUNE.

THE discovery of the planet Neptune took place under circumstances most extraordinary. The existence
of the planet was predicted, its path in the heavens was
assigned, its mass was calculated, from considerations
purely theoretical. The astronomer was told where to
direct his telescope, and he would see a planet hitherto
unobserved. The telescope was pointed, and there the
planet was found. In the whole history of astronomy
we can find few things equally wonderful. This discovery resulted from the study of the motions of the
planet Uranus.

Uranus was first discovered to be a planet in 1781,
but it had been repeatedly observed before by different
astronomers, and mistaken for a fixed star. Nineteen
observations of this description are on record, one of
them dating as far back as 1690. In 1821, M. Bouvard,
of Paris, published a set of tables for computing the

1*

place of this planet.   The materials for the construction
of these tables consisted of forty years' regular observations at Greenwich and Paris since 1781, and the
nineteen accidental observations, reaching back almost
a century further.   Upon comparing these observations,
Bouvard found unexpected difficulties.   He was unable
to find any elliptic orbit, which, combined with the
perturbations by Jupiter and Saturn, would represent
both the ancient and the modern observations.   When
he attempted to unite the ancient with the modern
observations, the former might be tolerably well represented, but the latter exhibited discordances too great
to be ascribed to errors of observation.   Not being
able to explain this discrepancy in any satisfactory manner, he rejected the ancient observations, and founded
his tables upon the observations since 1781.   " It being
necessary," says he, " to decide between the ancient and
the modern observations, I have held to the modern
ones as being the most likely to be accurate, and I leave
it to time to show whether the difficulty of reconciling
the two sets of observations depends upon the inaccuracy of the ancient ones, or on *some foreign and unknown influence to which the planet is subjected.*"
These tables represent very well the observations of
the forty years from which they were derived; but
soon after 1821, new discrepancies began to appear,
which have lately increased with great rapidity.   In
1832, the discordance between the observed and computed place of the planet, amounted to nearly half a

minute of space, and now the error exceeds two minutes. In order to exhibit more palpably the nature of these discrepancies, I have represented them upon the figure on the next page. If the straight line, A B, be taken to represent the path of Uranus, as computed from the elements of Bouvard, the broken line will represent the observed orbit. The deviation of these two lines indicates the discrepancy between theory and observation for the dates at the top of the page, the amount of the discrepancy being given on the left margin. If it was doubtful whether this difference in the case of the ancient observations was not due to the carelessness of the observers, no such supposition is admissible in the case of the observations since 1821. The discrepancy between Bouvard's orbit and the observations is now enormous, and is increasing with alarming rapidity.

What can be the cause of these discrepancies? Do they indicate errors in the computations of M. Bouvard? In order to decide this question, the illustrious Bessel, about the year 1840, subjected all the observations to a new calculation; and although he detected one or two errors of Bouvard, they did not materially influence the results. He satisfied himself that the ancient and modern observations could not be reconciled by any modification of the elements, and that the differences could not be attributed to inaccuracy of instruments, or to methods of observation. Are these anomalies due to the attraction of some unknown disturb-

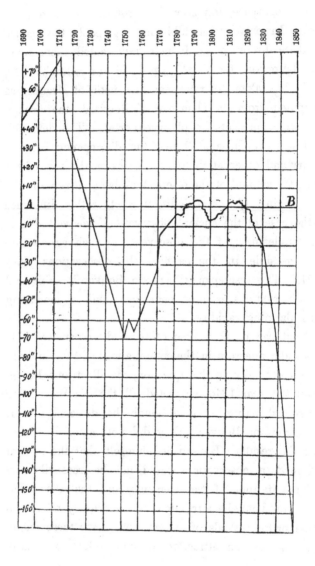

ing body? This idea was seriously entertained more than twelve years ago by Bouvard, Hansen, Hussey, Bessel, and some others. In a public lecture delivered in the year 1840, Bessel stated, "I have arrived at the full conviction that we have in Uranus a case to which Laplace's assertion, that the law of gravitation entirely explains all the motions observed in our solar system, is inapplicable. We have here to do with discordances whose explanation can only be found in a new physical discovery. Further attempts to explain them must be based upon the endeavor to discover an orbit and a mass for some unknown planet, of such a nature, that the resulting perturbations of Uranus may reconcile the present want of harmony in the observations."

Dr. T. J. Hussey, in 1834, proposed to compute an approximate place of the supposed disturbing body, and then commence searching for it with his large reflector. Mr. Airy, now Astronomer Royal of Great Britain, at that time professor in Cambridge, pronounced the problem hopeless. His words were: "If it were certain that there was any extraneous action upon Uranus, I doubt much the possibility of determining the place of the planet which produced it. *I am sure it could not be done till the nature of the irregularity was well determined from several successive revolutions;*" that is, till after the lapse of several centuries.

This deliberate opinion from one who, by common consent, stood at the head of British mathematicians and astronomers, would have deterred any but the most

daring mathematician from attacking the problem.
Again, in 1837, Mr. Airy repeats the same idea: "If
these errors are the effect of any unseen body, *it will
be nearly impossible ever to find out its place.*"

In the year 1842, the Royal Society of Sciences of
Göttingen, proposed, as a prize question, the full discussion of the theory of the motions of Uranus, with
special reference to the cause of the large and increasing error of Bouvard's tables. During the same year,
1842, Bessel was engaged in researches relative to this
problem; but his labors were soon interrupted by sick-
ness and subsequent death; and from this time we find
but two mathematicians, Mr. Adams, of Cambridge
University, in England, and M. Le Verrier, of Paris, who
busied themselves with the problem.

It should be remembered, that in accordance with the
Newtonian law of gravitation, every body in the solar
system attracts every other; that the attraction of each
body is proportioned to its quantity of matter; and that
in the same body the power of attraction varies inversely as the square of the distance. In order, therefore, to compute the exact place of a planet in its orbit
about the sun, it is necessary not merely to regard the
attraction of the central body, but also to allow for the

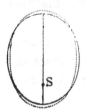

influence of all the other bodies of the
solar system. For instance, if the sun
alone had acted, Jupiter would revolve
around the central luminary in a perfect
ellipse, which we may represent by the
annexed black curve; but the real orbit

vibrates around this curve as in the dotted line, a curve which is not reproduced at every revolution, but will pass through an indefinite number of variations. To compute the exact orbit which a planet will describe, subject to the attractions of all the members of the solar system, is one of the grandest problems in astronomy.

Hitherto mathematicians had only aspired to compute the disturbing influence of one body upon another, when the magnitude and position of both bodies were known. But, in the case of Uranus, it was necessary to solve the inverse problem which Professor Airy had pronounced hopeless, viz., from the observed disturbances of one body, to compute the place of the disturbing body.

After taking his degree of Bachelor of Arts in January, 1843, with the honor of Senior Wrangler, Mr. Adams ventured to attack this problem, and obtained an approximate solution by supposing the disturbing body to move in a circle at twice the distance of Uranus from the sun. His results were so far satisfactory, as to encourage him to attempt a more complete solution. Accordingly, in February, 1844, having obtained through Professor Airy a complete copy of the Greenwich observations of Uranus, from 1754 to 1830, he renewed his computations, which he continued during that and the subsequent years. In September, 1845, he had obtained the approximate orbit of the disturbing planet, which he showed to Professor Challis, the director of the observatory at Cambridge; and near the close of the next month, he communicated his results to the

Astronomer Royal, together with a comparison of his theory with the observations. The discrepancies were quite small, except for the single observation of 1690, which differed by 44″; this observation not having been used in the equations of condition. Professor Airy, in acknowledging the receipt of this letter, pronounced the results extremely satisfactory, and inquired of Mr. Adams whether his theory would explain the error of the tables in regard to the distance of Uranus from the sun, which error he had shown to be very great. To this inquiry Mr. Adams returned no answer for nearly a year; probably because he was not able to answer the question entirely to his own satisfaction.

Meanwhile, this grand problem was undertaken by another mathematician who was entirely ignorant of the progress which Mr. Adams had made; for none of his results had yet been published. In the summer of 1845, M. Arago, of Paris, requested M. Le Verrier, a young mathematician, who had already distinguished himself by his improved tables of Mercury, to attempt the solution of this problem. This he accordingly did, and his success astonished all Europe. He commenced his investigations by inquiring whether the observations of Uranus could be reconciled with the supposition, that this body is subject to no other attraction than that of the sun and the known planets, acting according to the Newtonian law of gravitation. He carefully computed the effects due to the action of Jupiter and Saturn, neglecting no quantities until he had proved that their influence was insensible. He thus discovered some important terms which had been neglected by Laplace. He then compared his theory with observation, and proved conclusively that the observations of Uranus could not be reconciled with the law of gravitation, except by admitting some extraneous action. These results were communicated to the Academy of Sciences, Nov. 10, 1845; and such was the reputation secured by this and his preceding memoirs, that in January, 1846, he was elected to fill the vacancy which had occurred in the Institute in the section of Astronomy, by the death of Cassini. This memoir was but preliminary to his grand investigation; and it should be remarked, that Mr. Adams *had already deposited* with the Astronomer Royal at Greenwich, a paper containing the elements of the supposed disturbing planet, and agreeing closely with the results which Le Verrier subsequently obtained.

Le Verrier next proceeds to inquire after the cause of the discovered irregularities. Is it possible that at the immense distance of Uranus from the sun, the force of attraction does not vary inversely as the square of the distance? The law of gravitation is too firmly established to permit such a supposition, until every other resource has failed. Are these irregularities due to the resistance of a rare ether diffused everywhere through space? No other planet has afforded any indication of such a resistance. Can they be ascribed to a great satellite accompanying the planet? Such a cause would produce inequalities having a very short period; while the observed anomalies of Uranus are precisely the reverse. Moreover, it would be necessary to assign it such dimensions that it could not fail to have been visible in our telescopes. Has a comet impinged upon Uranus, and changed the form of its orbit? Such a cause might render it impossible to represent the entire series of observations by a single elliptic orbit; but the observations *before* the supposed collision, would all be consistent with each other, and the observations *after* collision would also be consistent with each other. Yet the observations of Uranus from 1781 to 1821, as may be seen from the diagram, page 12, accord neither with the earlier observations nor with the more recent ones.

There seems to remain no other probable supposition than that of an undiscovered planet. But if these disturbances are due to such a body, we can not suppose it situated within the orbit of Saturn. This would disturb the orbit of Saturn more than that of Uranus, while we know that its influence on Saturn is inappreciable, for Saturn's motion is well represented by the tables. Can this body be situated between Saturn and Uranus? We must then place it much nearer Uranus than Saturn, for the reason already assigned, in which case its mass must be supposed to be small, or it would produce too great an effect upon Uranus. Under these circumstances, its action would only be appreciable when in the immediate neighborhood of Uranus, which supposition does not accord well with the observations. The disturbing body must then be situated beyond Uranus, and at a considerable distance from it, for reasons already given. Now the distance of each of the more remote planets from the sun, is about double that of the preceding one. It is natural, then, to conjecture that the disturbing planet may be at a distance from the sun double that of Uranus, and it must move nearly in the ecliptic, because the observed inequalities of Uranus are chiefly in the direction of the ecliptic. Le Verrier then propounds the following specific problem :

"*Are the irregularities in the motion of Uranus due to the action of a planet situated in the ecliptic, at a distance from the sun double that of Uranus? If so, what is its present place, its mass, and the elements of its orbit?*" This problem he proceeds to resolve.

If we could determine for each day the precise effect produced by the unknown body, we could deduce from it the *direction* in which Uranus is drawn; that is, we should know the direction of the disturbing body. But the problem is far from being thus simple. The amount of the disturbance can not be deduced directly from the observations, unless we know the exact orbit which Uranus would describe, provided it were *free* from this disturbing action; and this orbit in turn can not be computed unless we know the amount of the disturbance. Le Verrier therefore computes for every nine degrees of the entire circumference, the effect which would be produced by supposing a planet situated in different parts of the ecliptic. He finds that when he locates the supposed disturbing planet in one part of the ecliptic, the discrepancies between the observed and computed effects are enormous. By varying the place of the planet, the discrepancies become smaller, until at a certain point they nearly disappear. Hence he concludes that there is but one point of the ecliptic where the planet can be placed, so as to satisfy the observations of Uranus. Having thus determined its approximate place, he proceeds to compute more rigorously its effects; and on the first of June, 1846, he announces as the result of his investigations, that the longitude of the disturbing planet for the beginning of 1847, must be about 325°.

The result thus obtained by Le Verrier, differed but one degree from that communicated by Mr. Adams to Professor Airy, more than seven months previous. Upon receiving this intelligence, Professor Airy expressed himself satisfied with regard to the general accuracy of both computations, and immediately wrote to Le Verrier, inquiring, as he had done before of Mr. Adams, whether his theory explained the error of the tables in respect to the distance of Uranus from the sun. Le Verrier answered that the errors of radius-vector *must* be accounted for, inasmuch as the equations of condition depended on observations at the quadratures as well as at the oppositions. Professor Airy was now so well convinced of the existence of a planet yet undiscovered, that he was anxious to have a systematic search for it forthwith undertaken. The observatory of Cambridge is provided with one of the finest telescopes of Europe, presented by the late Duke of Northumberland. Professor Airy urged upon the director, Professor Challis, to undertake the desired search; and recommended the examination of a belt of the heavens ten degrees in breadth, and extending thirty degrees in the direction of the ecliptic. This belt was to be swept over at least three times. If any star in the second sweep had a different position from that observed in the first, it might be presumed that it was the planet. If two sweeps failed of detecting the planet, it might be caught in the third.

Professor Challis commenced his search July 29th, and continued it each favorable evening, recording the exact position of every star down to the eleventh magnitude. Meanwhile Le Verrier was proceeding with his computations, and on the 31st of August he announced to the Academy the elements he had obtained for the supposed planet. He assigned its exact place in the heavens, and estimated that it should appear as a star of the eighth magnitude, with an apparent diameter of about three seconds; and, consequently, that the planet ought to be visible in good telescopes, and with a perceptible disc. The fixed stars are situated at such immense distances from us, that to the most powerful telescopes they appear only as *points*, although with a brilliancy augmented in proportion to the size of the telescope; while all the planets exhibit a measurable disc.

Le Verrier was of opinion that the new planet might be found by its possession of a visible disc, and therefore without any very great labor.

Soon after this communication was made to the Academy, Le Verrier, in acknowledging the receipt of a memoir, made use of the opportunity thus afforded to request Dr. Galle, of the Berlin observatory (where is found one of the largest telescopes of Europe), to undertake a search for his computed planet, and he assigned its supposed place in the heavens. The Berlin Academy had just published a chart of this part of the heavens, indicating the exact place of every star down to the tenth magnitude. On the evening of the very day upon which this letter was received (September 23), Galle found near the place computed by Le Verrier, a star of the eighth magnitude, not contained on the Berlin chart. Its place was carefully measured ; and the observations being repeated on the succeeding evening, showed a motion of more than a minute of space. The new star was found in longitude 325° 52′ ; the place of the planet computed by Le Verrier was 324° 58′ ; so that this body was within one degree of the computed point. Its diameter measured nearly three seconds. A coincidence so exact, left no doubt that this was really the body whose effects had been detected in the motions of Uranus. Mr. Galle accordingly writes to Le Verrier, "*the planet whose position you marked out actually exists.*" The news of the discovery spread rapidly over Europe. The planet was observed at Göttingen on the 27th of

September, at Altona and Hamburg on the 28th, and at London on the 30th.

We must now return to Professor Challis, whom we left exploring a large zone of the heavens, and recording the exact position of every star down to the eleventh magnitude. These observations were continued from the 29th of July to the 29th of September, during which time he had made more than three thousand observations of stars. On the 29th of September, Professor Challis saw for the first time Le Verrier's memoir communicated to the Academy August 31st. Struck with the confidence which Le Verrier manifested in his own conclusions, Professor Challis immediately changed his mode of observation, and endeavored to distinguish the planet from the fixed stars by means of its disc. On the same evening he swept over the zone marked out by Le Verrier, paying particular attention to the physical appearance of the brighter stars. Out of three hundred stars, whose positions were recorded that night, he selected one which appeared to have a disc, and which proved to be the planet. On the first of October he heard of the discovery at Berlin; and now, on comparing his numerous observations, he finds that he had *twice* observed the planet before, viz., on August 4th and 12th; but he lost the opportunity of being first to announce the discovery, by deferring too long the discussion of his observations.

The news of this capital discovery was brought to this country by the steamer of October 4th, and every telescope was immediately turned upon the planet. It was observed at Cambridge by Mr. Bond, October 21st; it was seen at Washington October 23d, and was regularly observed there till January 27th, when it approached too near the sun to be longer followed.

Le Verrier, although quite a young man, thus established at once an enviable reputation. He was literally overwhelmed with honors received from the sovereigns and academies of Europe. He was created an officer of the Legion of Honor by the King of France, and a special chair of Celestial Mechanics was established for him at the Faculty of Sciences. From the King of Denmark he received the title of Commander of the Royal Order of Dannebroga; and the Royal Society of London conferred on him the Copley medal. The Academy of St. Petersburg resolved to offer him the first vacancy in their body; and the Royal Society of Göttingen elected him to the rank of Foreign Associate.

Thus were the predictions of Adams and Le Verrier, with regard to the direction of the planet, at the present time, wonderfully fulfilled; but the observations of a few weeks sufficed to show that this body was not pursuing the orbit which these mathematicians had prescribed for it. Elements founded upon the hypothesis of a circular orbit were computed within the first month by Adams, Galle, and Binet. These agreed very nearly with one another, and coincided especially in showing the distance from the sun to be about 30. M. Valz, of the Marseilles observatory, endeavored, early in the year 1847,

to deduce the form of the orbit from the small arc
described by the planet since its discovery, but he found
himself unable to obtain a reliable value for the eccentricity. It became now a question of the greatest interest to determine with precision the elliptic elements
of the planet. This, however, was an object of unusual
difficulty, on account of the slow motion of the planet,
its heliocentric motion amounting to only about two
degrees in a year. The result derived from observations
embraced within a short interval of time must therefore
be received with great distrust, since a slight error in
the measurement of a minute portion of the orbit leads
to a much larger error in the computed length of the
remainder of the path. To furnish the orbit with much
precision, we must have observations extending over a
long series of years. Under these circumstances, it became a question of the highest interest, whether this
body may not have been observed by astronomers of
former years, and mistaken for a fixed star. If we could
obtain one good observation, made some time in the
last century, it would enable us at once to determine the
orbit with nearly the same precision as that of Jupiter
itself. It will then be presumed that astronomers have
not neglected to explore the records of the past, to discover if possible some chance observation of the new
planet.

Mr. Hind, of London, adopting the predicted elements
of Le Verrier, examined Lalande's and other observations
for this purpose; and satisfied himself that the new planet

was not there to be found. An American astronomer, Mr. Sears C. Walker, was more fortunate. Mr. Walker proceeded in the following manner. He first computed the orbit which best represented all the observations which had been made at the Washington observatory, as well as those which had been received from Europe. He then computed the planet's probable place for a long series of preceding years, and sought among the records of astronomers for observations of stars in the neighborhood of the computed path. Bradley, Mayer, and Lacaille have left us an immense collection of observations, yet they seldom recorded stars so small as the body in question. Among the observations of Piazzi, no one was found which could be identified with the planet. The Madras observations were generally confined to the stars of Piazzi's catalogue. The Paramatta catalogue seldom extends north of the thirty-third parallel of south declination ; and Bessel, in preparing his zones of 75,000 stars, did not sweep far enough south to comprehend the planet. The only remaining chance of finding an observation of the planet was among the observations of Lalande. The Histoire Celeste Française embraces 50,000 stars, and Mr. Walker soon found that Lalande had swept over the supposed path of the planet on the 8th and 10th of May, 1795. He accordingly computed more carefully the place of the planet for this period, making small variations in the elements of the orbit, so as to include the entire region within which the planet could possibly have been comprised. He then selected from the Histoire Celeste all the stars within a degree of the computed path. These stars were *nine* in number, of which *six* had, however, been subsequently observed by Bessel, and of course were to be set down as *fixed stars*. But three stars remained which required special examination; and of these, one was too small to be mistaken for the planet, and a second was thought to be too far from the computed place. The remaining star was distant only two minutes from the computed place of the planet; it was of the same magnitude, and was not to be found in Bessel's observations, although this part of the heavens must have been included in the field of his telescope. This discovery was made on the 2d of February, 1847; and on the first clear subsequent evening, February 4th, the great equatorial of the Washington observatory was pointed to the heavens, and *this star was missing*. Where Lalande, in 1795, saw a star of the ninth magnitude, there remained *only a blank*. The conclusion seemed almost certain, that Mr. Walker had here obtained the object of his search. He accordingly computed the path upon this supposition, and found that a single elliptic orbit would represent, with almost mathematical precision, the observation of 1795, and all the observations of 1846.

The case seemed completely made out. *But there was a weak point in the argument*. Lalande had marked his observation of the altitude of this star as *doubtful*. Could we rest the decision of a question so important upon a bad observation? How unfortunate, that among the 50,000

stars contained in this precious collection, there was
only one which could be presumed to have been the
planet, and this observation the author had marked as
doubtful! Thus the question stood—astronomers were
afraid to admit, and still could not reject the conclusions
of Mr. Walker. The steamer which left Boston on the
1st of March, carried a copy of the Boston Courier, containing the account of Mr. Walker's researches. This
paper was destined for M. Le Verrier; and, on the very
day of its arrival, he also received a letter from Altona,
dated March 21st, announcing that M. Petersen had discovered that this very star, observed by Lalande in
1795, was *now missing* from the heavens. M. Petersen's
discovery was made on the 17th of March. Mr. Walker
made the same discovery, theoretically, February 2d;
and it was confirmed by an actual inspection of the
heavens, February 4th. Mr. Walker, then, has the priority of *six weeks* in the discovery. Fortunately, the
original manuscripts of Lalande had been preserved, and
were deposited in the observatory of Paris. On con-
sulting them it was found that the doubtful mark ap-
pended to the published observations, did not exist in
the manuscript. Moreover, the star had been twice observed, viz., on the 8th and 10th of May, 1795; but as
the two observations did not agree, Lalande suppressed
the former, and in his printed book marked the latter
doubtful. The discrepancy between the two observations is almost exactly that which is due to two days'

motion of the planet, according to the orbit of Mr. Walker.

Thus, then, we have most unexpectedly secured *two good observations* in place of *one doubtful one.* We can no longer withhold our full belief. A single elliptic orbit represents with great precision the two observations of Lalande, and all the observations subsequently made.

More recently it has been announced that Dr. Lamont, of Munich, had twice observed the planet Neptune as a fixed star in his zones; the first time October 25th, 1845, when he estimated it as of the ninth magnitude; the second time was September 7th, 1846, when it was entered as of the eighth magnitude. The following, then, are the dates of the seven earliest observations of this planet, so far as at present known :

1795,	May 8th,		by Lalande.
1795,	" 10th,		by Lalande.
1845,	Oct. 25th,		by Lamont.
1846,	Aug. 4th,		by Challis.
1846,	" 12th,		by Challis.
1846,	Sept. 7th,		by Lamont.
1846,	" 23d,		by Galle.

Let us now compare the predicted orbits of Adams and Le Verrier with the true orbit according to Mr. Walker, and the mass of the planet as deduced from

Lassell's observations of the satellite. The comparison
stands as follows:

	ADAMS.	LE VERRIER.	WALKER.
Long. of the perihelion,	299° 11'	284° 45'	47° 14' 37"
Long. of ascending node,	*unknown.*	156 0	130 6 51
Inclination of the orbit,	*unknown.*	6 0	1 46 59
Mean long., Jan. 1, 1847,	323 24	318 47	328 32 44
True long., Jan. 1, 1847,	329 57	326 32	327 33 47
Eccentricity,	0.120615	0.10761	0.00871946
Mean distance from sun,	37.25	36.154	30.03666
Time of revolution in yrs.,	227.323	217.387	164.6181
Mean daily motion,	15".609	16".318	21".55448
Mass,	$\frac{1}{6666}$	$\frac{1}{9300}$	$\frac{1}{17500}$

In order to render the comparison more striking, I
have represented on the annexed figure the orbits of

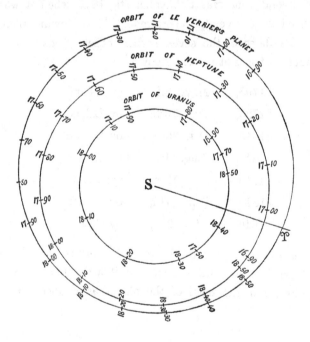

Le Verrier and Walker. The orbits of Adams and Le Verrier are almost identical, but both differ materially from Mr. Walker; that is, we are compelled to admit that they differ considerably from the truth. They represent remarkably well the *direction* in which the planet is now seen from the earth, but they give its distance too great by *three hundred millions of miles ;* and in 1690, the planet Neptune was more than *four thousand millions of miles* distant from the place assigned by Le Verrier to his planet, and differed nearly a quadrant from it in direction. This discrepancy is so great as to have given occasion for the remark, that the planet actually discovered is *not the planet predicted by Le Verrier.* What reason there may be for this remark I shall consider hereafter.

Some difficulty at first occurred in deciding upon a name for the new planet. The Bureau des Longitudes of Paris were in favor of calling it *Neptune,* and this name was given out by Le Verrier in private letters to different astronomers in England and Germany. Subsequently, Le Verrier commissioned his friend Arago to give the planet a name; and Arago declared he would never call it by any other name than *Le Verrier.* When Sir William Herschel discovered a planet, he named it Georgium Sidus; and the name of "the Georgian" was retained until recently in the English Nautical Almanac. But this name being offensive to the national pride of the French, they at first called the planet Herschel, and afterward Uranus. The latter name has now come into exclusive use; but Arago, in order to secure an honor to his friend Le Verrier, proposed to restore the name Herschel, and also that each of the smaller planets should receive the name of its discoverer.

The astronomers of Europe refused to concur in the decision of Arago. There are objections to this principle of nomenclature, some of which have considerable weight. The name of the discoverer of a planet may happen to be immoderately long, or ludicrously short, difficult to pronounce, or comically significant. Then, also, if the same astronomer should be fortunate enough to discover more than one planet, we should be obliged to repeat the surname with a prefix. Already we have two planets discovered by Olbers; two discovered by Hencke, ten by Hind, seven by Gasparis, five by Luther, four by Goldschmidt, and five by Chacornac.

Moreover, it often happens that several persons contribute an important part in the discovery of the same body. Thus the planet Ceres was first discovered by Piazzi, in the course of a series of observations having a different object in view. After a few weeks, the planet became invisible from its proximity to the sun. Astronomers computed the orbit from Piazzi's observations, and searched for it some months afterward, when it ought again to have come into view. But the planet could not be found. Ceres was entirely lost, and would not have been seen again, had not Gauss, by methods of his own invention, computed a much more accurate orbit, which disclosed the exact place of the fugitive, and enabled

De Zach to find it immediately upon pointing his telescope to the heavens. To Gauss, therefore, belongs the honor of being the *second discoverer* of Ceres; and the second discovery was far more glorious than the first.

The discovery of the new planet has been justly characterized by Professor Airy as " *the effect of a movement of the age.*" The honor of the discovery is not to be exclusively engrossed by either Adams or Le Verrier. The labors of numerous astronomers had prepared the way, and contributed more or less directly to the discovery. An eminent critic has ridiculed this idea. But Mr. Adams himself informs us, that his attention was first directed to the subject of the motions of Uranus, by reading Airy's report on the recent progress of astronomy; and Le Verrier states, that in the summer of 1845, he suspended the researches on comets, upon which he was then employed, to devote his time to Uranus, *at the urgent solicitation of M. Arago.* Omitting several who have indirectly contributed to this result, we find four whose names will ever be honorably associated with the discovery of the planet Neptune, viz., Adams, Challis, Le Verrier, and Galle. Adams first determined the approximate place of the new planet from the perturbations of Uranus. Professor Challis was the first to institute a systematic search for the planet, and had actually secured two observations of it before it was seen at Berlin. True, he did not at the time know that he had found the planet, for he had not interrogated his observations. But the prize was secured, and he would

2*

infallibly have recognized it as soon as he had instituted a comparison of his observations. Being fully resolved to make sure of the diamond, he shoveled up with it a great mass of rubbish, and stored it all away to examine at his leisure.

To Le Verrier belongs the credit of having· been the first to publish to the world the process by which he arrived at the conclusion of the existence of a new planet; and it is conceded that his researches were more complete and elaborate than those of his rival; while to Galle belongs the undisputed honor of having been the first practically to recognize this body as a planet.

To give to the new planet the name of Le Verrier, would be indeed to confer honor where honor was due; but it would be dishonor to others whose pretensions are but little inferior to his own. The astronomers of Europe have preferred to take a name from the divinities of the Roman mythology, in conformity with a well-established usage; and as the name of Neptune harmonizes with this system, and withal was first suggested by the Bureau des Longitudes, they decided to adhere to it. This was the unanimous voice of Europe, with the exception of France, and the astronomers of France have since acquiesced in this decision.

The discovery of Neptune has given an unequivocal refutation to Bode's law of the planetary distances. This famous law may be thus stated. If we set down the number 4 several times in a row, and to the second 4 add 3, to the third 4 add twice 3 or 6, to the next 4 add twice 6 or 12, and so on, as in the following table, the resulting numbers will represent nearly the relative distances of the planets from the sun:

$$
\begin{array}{cccccc}
4 & 4 & 4 & 4 & 4 & 4,\text{ etc.} \\
 & 3 & 6 & 12 & 24 & 48,\text{ etc.} \\
\hline
4 & 7 & 10 & 16 & 28 & 52,\text{ etc.}
\end{array}
$$

If the distance of the earth from the sun be called 10, then 4 will represent *nearly* the distance of Mercury; 7 that of Venus; and so of the rest. This law was never accurately verified in the case of any of the planets, and Neptune forms a decided exception to it. This fact is exhibited more clearly in the following table, which shows first the true relative distance of each of the planets; secondly, the distance according to Bode's law, and thirdly, the error of this law:

	True distance.	Bode's law.	Error.		True distance.	Bode's law.	Error.
Mercury,	3·87	4	0·13	Jupiter,	52·03	52	0·03
Venus,	7·23	7	0·23	Saturn,	95·39	100	4·61
Earth,	10·00	10		Uranus,	191·82	196	4·18
Mars,	15·24	16	0·76	Neptune,	300·36	388	87·64
40 Asteroids	26·09	28	1·91				

It will be seen from this table, that although this law represents pretty well the distances of the nearer planets, the error is quite large for Saturn and Uranus; and for Neptune the error is altogether overwhelming, amounting to more than *eight hundred* millions of miles, a quantity almost equal to the distance of Saturn from the sun. It is mere mockery to dignify such coincidences with the name of a *law*. A law of nature is *precise*—it is capable of exact numerical application. Let, then, the preceding rule be called the law of Bode; *it is not a law of nature*.

It was at first the opinion of some observers, that Neptune is surrounded by a ring like Saturn. Mr. Lassell, of Liverpool, has an excellent Newtonian reflector of twenty feet focal length, and two feet aperture, with which he has made numerous observations of the planet. On the 3d of October, 1846, he was struck with the shape of the planet, as being not that of a round ball; and again, on the 10th of October, he received a distinct impression that the planet was surrounded by an obliquely situated ring. On the 10th of November the planet appeared very much like Saturn, as seen with a small telescope, and low power, though much fainter. Several other persons also saw the supposed ring, and all in the same direction. During the season of 1847, Mr. Lassell frequently saw the same appearance again, and found its angle of position to be 70 degrees S. W. He also satisfied himself that this appearance did not arise from any defect in his telescope.

Professor Challis states that on the 12th of January, 1847, he received for the first time a distinct impression that the planet was surrounded by a ring. Two independent drawings, made by himself and his assistant, gave the annexed representation of its appearance. On the 14th, he saw the ring again, and was surprised that he had not noticed it in his earlier observations. The ratio of the diameter of the ring to that of the planet, was about that of three to two.

On the other hand, the great telescope at Cambridge, Mass., shows *no ring*. The following is the testimony of the director, Mr. W. C. Bond. "We are satisfied that there is not at present visible any ring surrounding Neptune within the reach of the Cambridge telescope. Both my son George and myself have repeatedly had opportunities of examining the planet under high powers with the full aperture of fifteen inches, and have seen only a round disc, while our micrometrical measures of his diameter agreed well together."

Mr. Lassell's present opinion appears to be less in favor of a ring than formerly. Yet, on the 29th of August, 1851, he recorded in his journal, "I received again an impression of a ring-like appendage, but it is principally on the south side, and is nearly at right angles to a parallel of declination." Also on the 4th of November, 1852, at Malta, in latitude 35° 53', under the most favorable atmospheric circumstances, he recorded in his journal, "I receive a decided impression of ellipticity in the direction of the greatest elongation of the satellite. There is an impression of an extremely flattened ring in the direction of the transverse axis. I think I have never seen Neptune so well before." It is understood that the astronomers at Pulkova have not yet succeeded in observing any appearance such as would lead to the suspicion of a ring. Under these circumstances, if we do not entirely deny the possible existence of a ring, we must at least hold our minds in suspense, and wait patiently for further evidence. It is possible that this question may not be fully cleared up until some more powerful telescope is turned upon the planet, or it can be seen in a different part of its orbit.

Neptune is attended by at least *one satellite*. Mr. Lassell states that on the 10th of October, 1846, he observed a faint star, distant from the planet about three diameters. On the 11th and 30th of November, and also December 3d, he saw a small star having about the same appearance; and he considered it probable that the star was a satellite. On the 7th of July, 1847, he again saw the supposed satellite, and on the following evening the planet and satellite had both changed their position with reference to the neighboring stars. On the 22d, 25th, and 26th of the same month, the planet appeared, attended by a satellite; and on the 1st of August he obtained the fullest evidence of the verity of the satellite, in being able clearly to ascertain, that during the two hours he watched the planet, it had carried the satellite along with it in its orbital motion. On the 20th of September, 1847, Mr. Lassell announced that during the current year he had obtained twenty observations of the satellite, and from them all he concluded that its time of revolution was five days, twenty hours, and fifty minutes. Mr. Bond, at Cambridge, during the years 1847 and 1848, obtained repeated measures of the distance and angle of position of this satellite by means of a micrometer with illuminated wires. By a combination of all the Cambridge observations, Mr. Bond has deduced the time of revolution five days and twenty-one hours. A discussion of the entire series of observations made by Mr. Lassell at Malta in 1852, gives the period five days, twenty-one hours, two minutes and forty-four seconds, and the motion of the satellite is found to be *retrograde*. The radius of its orbit is seventeen seconds, which gives about 236,000 miles for the distance of the satellite from the planet. The light of the satellite is nearly equivalent to that of a star of the fourteenth magnitude. It is always more brilliant in the S. W. than in the N. E. part of its orbit, presenting, in this respect, a striking analogy with the outer satellite of Saturn. In the former position, Mr. Lassell found it easy to observe, in the latter, extremely difficult.

APPARENT ORBIT OF THE SATELLITE OF NEPTUNE FOR 1852.

The preceding results enable us to compute the mass of Neptune, which is found to be $\frac{1}{17500}$ part of the sun.

Mr. Bond states that he has at times been quite confident of seeing a *second* satellite, but has never yet been able to obtain successive measures of its distance from the primary. Mr. Lassell also in August, 1850, suspected he saw a second satellite, but was unable to follow it long enough to establish its character.

The grand question still remains untouched— *Will the new planet explain the observed irregularities in the motion of Uranus ?*

The planet having been actually discovered in the heavens, by means of certain predicted elements, and within one degree of the predicted place, the natural conclusion was, that those elements were extremely near the truth, and that the planet would perfectly explain those effects by whose study its own existence had been detected. When, however, observation had rendered it certain that the planet moved in a smaller and less eccentric orbit than had been predicted, it became doubtful whether it would account for the anomalies in the motion of Uranus. When Le Verrier, in March, 1847, received notice of the computations of Mr. Walker, who obtained an orbit differing but little from a circle, he at once pronounced the small eccentricity *incompatible* with the observed perturbations. Mr. Adams, in a letter June 11th, 1847, says, "I am hard at work on the perturbations of Uranus, in order to obtain a new theoretical determination of the place. The general values of the perturbations are enormous, far exceeding any thing else of the same kind in the system of the primary planets. A comparison of the numerical expressions for the perturbations, which I have now obtained, with those which I used before, would *justify some skepticism as to former con-*

clusions. But we shall soon see how this great apparent difference affects the result."

By referring to the table on page 30, it will be seen that Adams and Le Verrier explain the anomalies of Uranus by assuming a very large body (having twice the mass of Uranus) moving in a very eccentric orbit. It is now discovered that the orbit hardly differs at all from a circle, and that the mass is scarcely one half of that which had been assumed. Now the more eccentric the orbit, the greater must be the *inequality* of a planet's action upon other bodies whose orbits are nearly circular. A considerable part of the observed irregularities in the motion of Uranus, was explained by Adams and Le Verrier, by means of the great eccentricity ascribed to their hypothetical planet. This portion is now gone with the failure of the eccentricity. But, on the other hand, Neptune is found to be much *nearer* Uranus than the hypothetical planet, and in consequence of this proximity, its disturbing action is *increased*, so that these two variations of the elements partly compensate each other. The mass of Neptune is also less than the hypothetical planet of Le Verrier, and on this account its disturbing action is diminished. To what extent the planet Neptune would account for the perturbations of Uranus was not determined until 1848. In a communication made to the American Academy April 4, 1848, Professor Peirce announced that the motions of Uranus are perfectly explained, provided we adopt Mr. Walker's orbit, and the mass of Neptune, which is derived from Mr. Bond's observations of Lassell's satellite. By his computations, the great anomalies which had been observed, and which are represented on page 12, almost entirely disappear. All the modern observations are represented quite as well as by the theories of Adams and Le Verrier, and the observation of 1690 much better. The error of Flamsteed's observation of 1690, according to Adams's computation, was 50″, and according to Le Verrier's computation, 20″; but according to Peirce's theory, this observation is represented within a single second.

So then the anomalies of Uranus, which had so long perplexed astronomers, are perfectly accounted for. But it is obvious, from a glance at the diagram on page 30, that the planet actually discovered is moving in an orbit considerably different from what had been computed, so that it has been claimed by Professor Peirce that *Neptune is not the planet whose existence had been predicted by Le Verrier.* Is this discrepancy between the observed and predicted orbits of a serious nature; and if so, how is it to be accounted for? This question has been fully discussed by Le Verrier himself. Le Verrier attempted to deduce the position of a new planet, by studying the irregularities in the motion of Uranus. The data which he was obliged to employ, were liable to some *uncertainty.* This uncertainty of the data did not result merely from the uncertainty of the *observations,* but from two other causes, viz., a possible error in the mass of Saturn sufficient to add 3″ to the uncertainty of the observations, and from the possible influence of a planet situated beyond

Neptune, whose action upon Uranus might easily amount
to 5″ or 7″. The uncertainty of the data results from the
combination of these three effects, and might amount to
10″ or 12″, while the uncertainty of the modern observations does not exceed 2″ or 3″. Now the irregularities in
the motion of Uranus, which serve as the basis of the discussion, do not exceed at the utmost two minutes of
space, and for the most part they are less than one
minute; so that the data were uncertain to *one tenth* of
their whole amount. Le Verrier, therefore, claims that it
is unreasonable to demand of him a greater degree of accuracy in the positions assigned to his computed planet,
than one tenth of their value, and he has attempted to
show that these positions are not in error by so large a
quantity. Moreover, as he computes the position of his
planet only by means of the disturbance which it causes
in the motion of Uranus, he can only compute this position when this disturbance is appreciable. Now this disturbance, on account of the distance of the planets from
each other, is inappreciable from 1690 to 1812. It was
only from 1812 to 1842, that he was furnished with observations in which the disturbing action of Neptune was
sensible, and the place of the planet for any other time
must be deduced from its motion during these 30 years.
He then proceeds to show that the positions he had assigned to Neptune during this period, and also for a
much longer time, were not in error by one tenth of their
whole amount. Let us consider,

I. The error in the computed longitudes of Neptune.

The following is the comparison between Le Verrier's predicted places and those resulting from Mr. Walker's orbit during a period of 120 years.

Year.	Predicted longitude.	Longitude according to Mr. Walker.	Error.
1767	174°·3	155°·5	+18·°8
1777	190·4	176·7	+13·7
1787	207·6	198·0	+ 9·6
1797	225·9	219·3	+ 6·6
1807	245·2	240·7	+ 4·5
1817	265·3	262·2	+ 3·1
1827	285·9	283·9	+ 2·0
1837	306·4	305·7	+ 0·7
1847	326·5	327·5	— 1·0
1857	345·7	349·7	— 4·0
1867	3·9	11·6	— 7·7
1877	20·9	33·6	—12·7
1887	36·9	55·5	—18·6

Thus, it appears, that during a period of 120 years, the longitude of Neptune, according to Le Verrier's computation, did not differ from that determined by Mr. Walker from the observations, more than 18°·8, or about *one twentieth* part of an entire circumference; and during the period in which the action of Neptune has been clearly marked, that is, since 1812, the error of Le Verrier's theory has not exceeded 3°·7. For 1690, the theory is indeed very much at fault, but that is the result of a comparatively small error in the positions assigned during the present century.

II. The error in the computed distance from the sun.

The following table shows the distance of the planet from the sun, as predicted by Le Verrier, and that deduced from the computations of Mr. Walker, the distance being expressed in radii of the earth's orbit.

	Predicted distance.	Observed distance.	Error.
In 1812	32·7	30·4	+ 2·3
1822	32·3	30·3	+ 2·0
1832	32·6	30·2	+ 2·4
1842	32·8	30·1	+ 2·7

Thus it appears, that during the 30 years in which the action of Neptune upon Uranus has been sensible, the error of the predicted distance from the sun has never amounted to *one tenth* of the whole quantity.

There appears, however, a discrepancy between the *limits* of distance which Le Verrier assigned to his planet before its discovery, and those which he has since published. In 1846 he attempted to determine the limits within which the distance might be supposed to vary without involving an error greater than 5″ in any of the observations since 1781. *He decided that the mean distance could not be less than* 35·04, *nor more than* 37·90. Now the mean distance of Neptune from the sun is known to be only 30·04. From these two propositions the legitimate conclusion would seem to be that Neptune is not the planet predicted by Le Verrier. But Le Verrier has lately changed his ground, and he has discovered that without supposing the uncertainty of the modern data to exceed 5″, the theory of Uranus may be satisfied by a planet situated in 1846 at any distance from

the sun between 29·6 and 35·2. It is not obvious how
this last statement can be reconciled with the limits
published in 1846. Doubtless Le Verrier has discovered
that the limits which he first assigned were erroneous.
If we compare the predicted *mean distance* with that
deduced from the observations, we shall find the error
of the former to amount to *one fifth* of the whole
quantity.

III. The error in the eccentricity of Neptune.

Le Verrier assigned to the orbit of his planet an eccentricity of 0·1076; the computations of Mr. Walker
make the eccentricity of Neptune 0·0087.

The discrepancy is considerable; but Le Verrier states
that if we admit an uncertainty of 5″ only in the modern
data, the eccentricity of the body producing the irregularities of Uranus may be chosen arbitrarily between
0·2031 and 0·0592; and if we admit an uncertainty of
7″ or 8″ in the modern data, the eccentricity may have
any value between 0·25 and *zero*. Professor Peirce has
shown that the planet Neptune, with an eccentricity
almost zero, reconciles all the modern observations of
Uranus within 3″.

IV. Error of the computed mass of Neptune.

The mass assigned by Le Verrier to his predicted
planet was $\frac{1}{9300}$ of the sun's mass; but he adds, that if
we admit an uncertainty of 5″ in the data, this mass may
have any value between $\frac{1}{4700}$ and $\frac{1}{14400}$. The mass of
Neptune deduced by Struve, from his own observations
of the satellite, is $\frac{1}{14194}$; but the mass deduced from Las-

sell's observations, is $\frac{1}{17500}$. The former value comes fairly within the limits assigned by Le Verrier, the latter somewhat exceeds them.

On the whole, we must conclude that the orbit of Neptune agrees with the orbit predicted by Le Verrier, very nearly within the limits which Le Verrier *now assigns*, upon the supposition of an uncertainty of 5″ in the data since 1781. But these limits, with respect to distance and time of revolution, are very different from those assigned before the discovery of the planet. The orbit of Neptune is not included within the limits which Le Verrier then assigned; and it is a legitimate inference, from his own premises, that Neptune is not the planet whose existence he announced.

It should also be borne in mind that Professor Peirce has shown that Neptune reconciles all the observations of Uranus since 1781 within 3″, and that the greatest error of any of the ancient observations according to his theory is 8″; thus proving that Le Verrier's suspicion of the existence of a planet beyond Neptune, and of an error in the mass of Saturn is unfounded. The data, therefore, instead of being uncertain to one tenth of their whole amount, were generally reliable within one *sixtieth* of their value; and Le Verrier's elements are erroneous to an extent far beyond the one sixtieth of their value.

It will naturally be asked, how has it happened that two astronomers, Adams and Le Verrier, have arrived, by independent computations, at almost identically the same result, and have made such mistakes with respect to the mean distance, eccentricity, and mass of the planet? The answer is plain: they were misled by placing too great confidence in Bode's law of the planetary distances. Since it was necessary, in the first instance, to make some hypothesis with regard to the distance of the disturbing body from the sun, both computers started with that supposition which was generally thought most probable. The distance of Saturn from the sun is nearly double that of Jupiter; the distance of Uranus is almost exactly double that of Saturn; hence it seemed probable that the planet they were in search of, would be found at a distance about double that of Uranus. Accordingly this assumption was made the basis of their first computations; but neither of the computers accepted this as his final result, without attempting to verify it. They both varied the assumed distance, and found that by bringing the planet a little nearer the sun, the observed irregularities of Uranus were still better explained. The distance of 36·154 (or about 3435 millions of miles), finally adopted by Le Verrier, was that which appeared to reconcile all the observations *most satisfactorily*. This distance corresponds to a period of two hundred and seventeen years. Le Verrier found (or thought he found) that whether he increased or diminished this distance, the observations of Uranus were not so well represented. He hence inferred that the mean distance from the sun could not be less than 35·04, nor greater than 37·90. (Le demi-grand axe de l'orbite *ne peut varier* qu'entre les limites 35·04 et 37·90.) The periods of revolution corresponding to these

distances are about 207 and 233 years. This conclusion
was unauthorized, and is now admitted by Le Verrier to
have been erroneous. The mean distance of Neptune
from the sun is only thirty times that of the earth, and
still it explains the motions of Uranus even better than
the hypothetical planet of Le Verrier.

Professor Peirce has shown that an important change
in the character of the perturbations takes place near the
distance 35·3. A planet at the distance 35·3 would revolve about the sun in 210 years, which is exactly two
and a half times the period of the revolution of Uranus.
Now if the times of revolution of two planets were exactly as 2 to 5, the effects of their mutual influence would
be peculiar and complicated. This distance of 35·3 is a
complete barrier to any logical deduction, and the investigations with regard to the outer space can not be extended to the interior.

The observed distance 30 belongs to a region which is
even more interesting in reference to Uranus than that of
35·3. The time of revolution which corresponds to the
mean distance 30·4 is 168 years, being exactly double the
year of Uranus; and the influence of a mass revolving in
this time would give rise to very singular and marked
irregularities in the motion of this planet. Professor
Peirce was hence led to the conclusion that the planet
Neptune was not the planet to which geometrical analysis
had directed the telescope; that its orbit was not contained within the limits of space explored by Adams and Le
Verrier in searching for the source of the disturbances of

3

Uranus; and that its discovery by Galle must be regarded
as a happy accident. Besides that solution of the problem which Le Verrier and Adams obtained, there is
another solution which corresponds to the orbit and mass
of Neptune. The fact however that Neptune does not
correspond to Le Verrier's solution can not detract from
the merit or value of his investigation. Since, by using
all the observations within his reach, he found an orbit
and mass capable of accounting for the observed motions
of Uranus, he is entitled in the opinion of mathematicians
to all the admiration he would have received had such a
planet actually moved in that orbit.

To some it has appeared a matter of surprise that the
new planet was not sooner discovered. Le Verrier's second memoir, which assigned the probable place of the
disturbing body, was presented to the Academy on the
first of June, 1846; and his third memoir (containing
every thing which Dr. Galle had in his possession at the
time of his discovery) was presented August 31st; yet
Galle's discovery was not made till September 23d.
What were the astronomers of Paris doing meanwhile?
Why did they not immediately point their telescopes to
the heavens? Why did they neglect the opportunity of
securing to France the glory of both the theoretical and
practical discovery, and leave to a German astronomer
the verification of the sublimest theory of modern
science? The answer is plain. *The astronomers of Paris
did not expect to find a planet within one degree of the place
computed by Le Verrier.* Le Verrier himself did not ex-

pect it. He assigned the *most probable* place of his planet in longitude 325°. He expressed the opinion that its longitude would not be *less* than 321°, nor *more* than 335°. But he adds, "If the planet should *not* be discovered within these limits, then *we must extend our search beyond them*." If he was sure of being able to find his planet without a long-continued and laborious search, why did he not borrow a telescope, and at once verify his own predictions?

Nor had the astronomers of the rest of Europe much higher faith than those of Paris. Professor Encke, in announcing the discovery, characterizes it as "*far exceeding any expectations which could have been previously entertained*." That Professors Airy and Challis, although they were pretty well satisfied of the *existence* of a planet yet undiscovered, regarded its exact place in the heavens as extremely uncertain, is plain from their comprehensive plan of observation, viz., to sweep three times over a belt of the heavens, thirty degrees in length, and ten degrees in breadth, a plan which Professor Challis states it would have been impossible for him to complete within the year 1846.

Do we then charge Encke and Airy with a want of sagacity? By no means. On the contrary, we maintain *that they had no reason to expect to find the planet within one degree of the computed place*. Le Verrier's own statement of the limits within which the planet should be sought for, is sufficient proof of this. "L'incertitude des données pourrait produire une incertitude de plus

de 18 dégrés dans le lieu de l'astre, à l'une des epoques
où l'on pouvait le mieux repondre de sa position.   The
uncertainty of the data caused *an uncertainty of more
than eighteen degrees in the position of the planet, even at the
time when its situation was best determined.*" Professor
Challis, therefore, proceeded like a sagacious as well as
brave general. He contemplated a long siege—yet his
plan rendered ultimate success almost certain.   Dr. Galle
took the citadel by storm—yet he had no reason to expect so easy a conquest. His success must have astonished himself as much as it did the world.

Let us then be candid, and claim for astronomy no
more than is reasonably due. When in 1846 Le Verrier
announced the existence of a planet hitherto unseen,
when he assigned its exact position in the heavens, and
declared that it shone like a star of the eighth magnitude,
and with a perceptible disc, not an astronomer of France,
and scarce an astronomer in Europe, had sufficient faith
in the prediction to prompt him to point his telescope
to the heavens. But when it was announced that the
planet had been seen at Berlin ; that it was found within
one degree of the computed place ; that it was indeed a
star of the eighth magnitude, and had a sensible disc,
then the enthusiasm not merely of the public generally,
but of astronomers also, was even more wonderful than
their former apathy. The sagacity of Le Verrier was
felt to be almost superhuman. Language could hardly
be found strong enough to express the general admiration. The praise then lavished upon Le Verrier was

somewhat extravagant. The singularly close agreement between the observed and computed places of the planet was accidental. So exact a coincidence could not have been reasonably anticipated. If the planet had been found even ten degrees from what Le Verrier assigned as its most probable place, this discrepancy would have surprised no astronomer. The discovery would still have been one of the most remarkable events in the history of astronomy, and Le Verrier would have merited all the honors which have since been conferred upon him.

SECTION II.

SEVENTY-FIVE years since, the only planets known to men of science were the same which were known to the Chaldean shepherds thousands of years ago. Between the orbit of Mars and that of Jupiter, there occurs an interval of no less than 350 millions of miles, in which no planet was known to exist before the commencement of the present century. Nearly three centuries ago, Kepler had pointed out something like a regular progression in the distances of the planets as far as Mars, which was broken in the case of Jupiter. Having despaired of reconciling the actual state of the planetary system with any theory he could form respecting it, he hazarded the conjecture that a planet really existed between the orbits of Mars and Jupiter, and that its smallness alone prevented it from being visible to astronomers. The remarkable passage containing this conjecture is found in his *Prodromus,* and is as follows: " When this plan, therefore, failed, I tried to reach my aim in another way, of, I must confess, singular boldness. Between Jupiter and Mars I interposed a new planet, and another also between Venus and Mercury, both which it is possible are not visible on account of their minuteness, and

I assigned to them their respective periods. In this way I thought that I might in some degree equalize their ratios, which ratios regularly diminished toward the sun, and enlarged toward the fixed stars."

But Kepler himself soon rejected this idea as improbable, and it does not appear to have received any favor from the astronomers of that time.

An astronomer of Florence, by the name of Sizzi, maintained, that as there were only seven apertures in the head—two eyes, two ears, two nostrils, and one mouth—and as there were only seven metals and seven days in the week, so there could be only seven planets. These seven planets, according to the ancient system of astronomy, were Saturn, Jupiter, Mars, the Sun, Venus, Mercury, and the Moon.

In 1772, Bode published a treatise on Astronomy, in which he first announced the singular relation between the mean distances of the planets from the sun, which has since been distinguished by his name. This famous law may be thus stated. If we set down the number four several times in a row, and to the second 4 add 3, to the third 4 add twice 3 or 6, to the next 4 add twice 6 or 12, and so on, as in the following table, the resulting numbers will represent nearly the relative distances of the planets from the sun:

4	4	4	4	4	4	4
	3	6	12	24	48	96
4	7	10	16	28	52	100

If the distance of the Earth from the Sun be called 10, then 4 will represent *nearly* the distance of Mercury; 7 that of Venus; 16 that of Mars; 52 that of Jupiter; and 100 that of Saturn. This law exhibited in a striking light the abrupt leap from Mars to Jupiter, and suggested the probability of a planet revolving in the intermediate region. This conjecture was rendered still more plausible by the discovery, in 1781, of the planet Uranus, whose distance from the sun was found to conform nearly with the law of Bode. In Germany, especially, a strong impression had been produced that a planet really existed between Mars and Jupiter, and the Baron de Zach went so far as to calculate, in 1784–5, the orbit of the ideal planet, the elements of which he published in the Berlin Almanac for 1789. In 1800, six astronomers, of whom the Baron was one, assembled at Lilienthal, and formed an association of twenty-four observers, having for its object to effect the discovery of the unseen body. For this purpose the zodiac was divided into twenty-four zones, one of which was to be explored by each astronomer; and the conduct of the whole operation was placed under the superintendence of Schröter. Soon after the formation of this society, the planet was discovered, but not by any of those astronomers who were engaged expressly in searching for it. Piazzi, the celebrated Italian astronomer, while engaged in constructing his great catalogue of stars, was induced carefully to examine, several nights in succession, a part of the constellation Taurus, in which Wollaston, by mistake, had assigned the position of a star which did not really exist. On the 1st of January, 1801, Piazzi observed a small star, which on the following evening appeared to have changed its place. On the 3d he repeated his observations, and he now felt assured that the star had a retrograde motion in the zodiac. On the 24th of January he transmitted an account of his discovery to Oriani and Bode, communicating the position of the star on the 3d and 23d of that month. He continued to observe the star until the 11th of February, when he was seized with a dangerous illness, which completely interrupted his labors. His letters to Oriani and Bode did not reach those astronomers until the latter end of March, at which time the planet had approached too near the sun to admit of further observations, and it was necessary for this purpose to wait until the month of September, when the planet would have extricated itself from the solar rays. Its re-discovery, after the lapse of so considerable a period, subsequent to the most recent observation, could not be accomplished without a pretty accurate knowledge of the orbit in which it was moving; but the data communicated by Piazzi were insufficient for this purpose. After some delay he communicated to astronomers all the observations made by himself down to the end of February. Professor Gauss found that they might all be satisfied within a few seconds by an elliptic orbit, of which he calculated the elements; and with the view of aiding astronomers in searching for the planet, he computed an ephemeris of its motion for

3*

several months. The planet was finally discovered by
De Zach on the 31st of December, and by Olbers on the
following evening. Piazzi conferred on it the name of
Ceres, in allusion to the titular goddess of Sicily, the
island in which it was discovered; and the sickle has
been appropriately chosen for its symbol of designation.

The mean distance of Ceres, as determined by the calculations of Gauss, was 2·767. The distance assigned by
Bode's law was 2·8. In this respect, therefore, the newly-
discovered planet harmonized with the other bodies of
the system to which it belonged. The new planet was,
however, excessively minute; its diameter, according to
Herschel's measurements, amounting to only 161 miles.
Its inclination to the ecliptic exceeded ten degrees, and
consequently it deviated from that plane more than either
of the older planets.

The discovery of Piazzi was soon followed by another
of a similar nature. Dr. Olbers, while engaged in
searching for Ceres, had studied with minute attention
the various configurations of all the small stars lying
near her path. On the 28th of March, 1802, after observing the planet, he swept over the north wing of
Virgo with an instrument termed a "Comet Seeker,"
and was astonished to find a star of the seventh magnitude, forming an equilateral triangle with two other small
stars, whose positions were given in Bode's catalogue,
where he was certain no star was visible in January and
February preceding. In the course of less than three
hours he found the right ascension had diminished, and

the north declination increased. On the following evening, as soon as the twilight permitted, he looked again for his star; it no longer formed an equilateral triangle with the stars above mentioned, but had moved considerably in the direction indicated by the preceding night's observations. On the 30th, after again observing the planet, Dr. Olbers wrote to Bode at Berlin, and to Baron De Zach, giving an account of his discovery. "What a singular accident," he exclaims, "was it by which I found this stranger nearly in the same place where I had observed Ceres on the 1st of January!" The elements of the orbit were quickly determined by Professor Gauss, who found the most remarkable peculiarity consisted in the great inclination of its plane to the ecliptic, which amounted to 34° 35'. The orbit was found to be an ellipse of not much greater eccentricity than that of Mercury, with a mean distance nearly *the same as that of Ceres*. Dr. Olbers suggested Pallas as the name for this new member of our system.

A comparison of the relative magnitudes of the planetary orbits had suggested the existence of an unknown planet, revolving between the orbits of Mars and Jupiter. Instead of one planet, however, two had been discovered. Olbers remarked that the orbits of these two bodies approached very near each other at the descending node of Pallas, and he conjectured that they might possibly be the fragments of a larger planet which had once revolved in the same region, and had been shivered in pieces by some tremendous catastrophe; and he intimated that there might be many more similar fragments which had not yet been discovered. He also inferred, that though the orbits of all these fragments might be differently inclined to the ecliptic, yet, as they all had a common origin, their orbits would have two common points of intersection, situated in opposite regions of the heavens, through which every fragment would necessarily pass in the course of each revolution. He proposed, therefore, to search carefully, every month, the north-western part of the constellation Virgo, and the western part of the constellation of the Whale, being the two opposite regions in which the orbits of Ceres and Pallas were found to intersect each other. Meanwhile the discovery of a third planet tended to confirm the truth of his hypothesis, and to encourage him in his arduous undertaking.

Professor Harding, of Lilienthal, undertook to construct a series of charts upon which should be represented the positions of all the small stars lying near the paths of Ceres and Pallas, with a view to assist the identification of these minute bodies. On the 1st of September, 1804, while engaged in exploring the heavens for this purpose, he perceived a small star in the constellation Pisces, very near to that part of the constellation of the Whale through which Olbers had asserted that the fragments of the shattered planet would be sure to pass. On the evening of the 4th he re-examined the neighborhood, and found that the star had changed its place. On the 5th and 6th, he observed it more accurately, and finding that the positions deduced from his observations confirmed the motion indicated by the estimates on September 1st and 4th, he announced the discovery to Dr. Olbers, at Bremen, on the 7th, who saw it the same evening. Pro fessor Harding named his planet Juno. The elements of its orbit were calculated by Gauss, who found its mean distance from the sun to coincide nearly with the mean distances of Ceres and Pallas. The eccentricity surpass- ed that of any other member of the planetary system. Like Ceres and Pallas, it is remarkable for its extreme smallness. Herschel was unable to pronounce with cer- tainty that its diameter exhibited any sensible magni- tude.

Stimulated by the discovery of Juno, Olbers continued with unremitting assiduity to explore the two opposite re- gions of the heavens through which he conceived the fragments of the shattered planet must pass. At length, after he had been engaged nearly three years in this la- borious pursuit, his perseverance was crowned with suc- cess. On the evening of the 29th of March, 1807, while occupied in sweeping over the north wing of Virgo, he discovered an object shining like a star of the sixth or seventh magnitude, which he concluded at once to be a planet, inasmuch as the previous examination of the vicinity had indicated no star in the position of the stranger. On the same evening he satisfied himself that it was really in motion, and continuing his observations until the 2d of April, he obtained sufficient evidence to justify the public announcement of his discovery of another new planet. Accordingly, on the following day,

he wrote to Professor Bode of Berlin, and to Baron de
Zach of Gotha, and particularly mentioned that his second discovery was not the result of accident, but of a
systematic search for a body of this nature.   The elements of the orbit were determined by Gauss, who executed the calculations required for this purpose within
ten hours after he obtained possession of the observations.
The planet was found to revolve in the same region with
Ceres, Pallas, and Juno, its mean distance from the sun
being somewhat less than that of either of those bodies.
At the request of Dr. Olbers, Gauss consented to name
the planet, and decided upon Vesta, the symbol of designation being the altar on which burned the sacred fire in
honor of the goddess.   This planet is even smaller than
either of the three others previously discovered, but it is
remarkable for the brilliancy of its light.   Near her opposition to the sun, a person with good sight may often
distinguish her without a telescope.

Dr. Olbers continued his systematic examinations of the
small stars in Virgo and Cetus, between the years 1808
and 1816, and was so closely on the watch for any moving body, that he considered it very improbable a planet
could have passed through either of these regions in the
interval, without being detected.   No further discovery
being made, the plan was relinquished in 1816.

In 1825, a fresh impulse was given to researches of
this nature, by the resolution of the Berlin Academy of
Sciences to procure the construction of a series of charts
representing the positions of all the stars, down to the

ninth magnitude, in a zone of the heavens extending fifteen degrees on each side of the equator. Only about two thirds of the charts contemplated in this great undertaking have yet been executed.

About the year 1830, M. Hencke, an amateur astronomer of Driessen, in Germany, commenced a careful survey of the zone of the heavens comprised within the charts published by the Academy of Berlin. He extended those maps by the insertion of smaller stars, and made himself well acquainted with their various configurations. After fifteen years his perseverance met with its due reward. On the 8th of December, 1845, while engaged in comparing the map of the fourth hour of right ascension with the heavens, he noticed what appeared to be a star of the ninth magnitude, between two others of the same brightness in Taurus, which had not been noted in his previous examinations. Without waiting for any further observations, M. Hencke wrote to Professors Encke and Schumacher, stating his reasons for supposing that he had detected a new planet. On the 14th of December the Berlin astronomers found the stranger in a position where no star was marked on the corresponding chart of the Academy, and the motion was easily perceived the same evening. On this occasion the elements of the orbit were rapidly determined, not by Gauss individually, as on previous occasions of a similar kind, but by a host of young astronomers throughout Europe, who had become familiar with the methods of that illustrious master. The results of their calculations showed the body to be one of the family of asteroids. M. Hencke requested Professor Encke to name his new planet, and the Professor conferred on it the appellation of Astræa.

Encouraged by his success, M. Hencke continued his search for planetary bodies, extending and verifying the Berlin Academical charts, and by frequent comparison with the heavens acquired an extensive knowledge of the configurations of the smaller stars in certain regions about the equator and ecliptic.   On the 1st of July, 1847, while engaged in examining the seventeenth hour of right ascension, he perceived a small star, of about the ninth magnitude, which was not marked on the corresponding map of the Academy.   On the 3d, he repeated his observation, and found that, during the intermediate period, its right ascension had sensibly diminished, leaving no doubt of its planetary nature.   Information of the discovery was circulated by M. Hencke on the following day, and the planet was soon recognized at the principal observatories of Europe.   The illustrious mathematician, Professor Gauss, was deputed by the discoverer to select a name for the stranger, and it received the name of Hebe, with a cup for the symbol, emblematic of the office of the goddess in mythology.   The orbit is very eccentric, and inclined more than 14 degrees to the plane of the ecliptic.

The next two members of this remarkable group in order of discovery were found by Mr. Hind, at the observatory erected by Mr. Bishop, in the grounds of his private residence in the Regent's Park, London. So early as April, 1845, a search for planetary bodies was commenced, but in consequence of other classes of observation, no systematic plan of examination of the heavens was attempted. In November, 1846, a rigorous search was undertaken, the Berlin Academical charts being employed as far as they extended; while ecliptic charts, including stars to the tenth magnitude, were formed for other parts of the heavens, where the ecliptic passes beyond the limits of the Berlin maps. On the 13th of August, 1847, after nine months' close observation on the above system, an object resembling a star of the eighth magnitude was discovered, which was not marked on the corresponding Berlin map. Its planetary nature being immediately suspected, it was attentively observed, and in less than half an hour the motion in right ascension was detected. In the course of an hour the planet had retrograded two seconds of time, a sufficient change of place to be indubitable. An announcement of the discovery was made to astronomers generally on the following morning, and observations were soon obtained at most of the European observatories. At the suggestion of Mr. Bishop the planet was named Iris. The symbol is due to Professor Schumacher, and is composed of a semicircle representing the rainbow, with an interior star, and a base line for the horizon. Several observers have remarked decided variations in the light of this planet, which are not accounted for by change of distance from the earth and sun, and which there is strong reason to suppose are in a great measure independent of atmospheric conditions.

Continuing the plan of observation already described, Mr. Hind noticed, on the 18th of October, 1847, in the constellation Orion, a star of the eighth or ninth magnitude, which had not been previously visible in the position it then occupied. Micrometrical measures of its position, made after the lapse of about four hours from the time when he first observed it, established the existence of a proper motion, and it was immediately announced to astronomers as the eighth member of the group of small planets. At the suggestion of Sir John Herschel the new planet received the name Flora, and a flower, the "rose of England," was chosen as the symbol. Its period of revolution is shorter than that of any other of the asteroids, being only about 1193 days. Flora, therefore, comes after Mars in order of mean distance from the sun, and approaches nearer to the earth than the rest of the group to which she belongs. The planet is somewhat ruddy, but without any hazy appearance, such as might be supposed to arise from an extensive atmosphere.

In the year 1848 another member of this interesting group was brought to light by Mr. Graham, at the private observatory of Markree Castle, Ireland, under the direction of Mr. Cooper. Having formed a chart of the stars near the equator, in the 14th hour of right ascension, on a more extended scale than that of the Berlin charts, he remarked, on the 25th of April, a star of the tenth magnitude in a position where none had been visible before, and noted it down for re-examination. On the following evening this object was found to have retrograded one minute, thus leaving no doubt of its planetary nature. On the 27th the discovery was announced to several astronomers in England and on the Continent, and soon became generally known through the circulars issued by Professor Schumacher. The name selected for this planet is Metis, with an eye and star for a symbol. This planet is remarkable for the near coincidence of its mean motion with that of Iris, the difference of their periodic times, according to the most recent calculations, amounting to less than one day.

On the 12th of April, 1849, Dr. Annibal de Gasparis, assistant astronomer at the royal observatory at Naples, while comparing the Berlin chart for the twelfth hour of right ascension with the heavens, perceived a star of between the ninth and tenth magnitudes, in a position which he had found vacant at previous examinations of this region. Unfavorable weather interrupted his observations for that evening, but on the 14th he ascertained that it had sensibly changed its place, and was therefore a new planet. Professor Capocci, director of the Neapolitan observatory, named the planet Hygeia. The mean distance of this planet from the sun is, with perhaps a single exception, greater than that of any other known member of this group, corresponding to a revolution in 2041 days.

On the occasion of the discovery of Hygeia, Sir John

Herschel had suggested that Parthenope would be a very appropriate name to commemorate the site of the discovery; that nymph having given her name to the city now called Naples. Signor de Gasparis states that he used his utmost exertions to realize for Sir John Herschel a Parthenope in the heavens, and his endeavors were crowned with success on the 11th of May, 1850. This new planet appeared like a star of the ninth magnitude.

On the evening of September 13, 1850, Mr. Hind noticed a star of the eighth magnitude in the constellation Pegasus, near another small one frequently examined on previous occasions, without any mention being made of its bright neighbor. Its peculiar bluish light satisfied him at once as to its planetary nature, and the micrometer was introduced to ascertain the difference of right ascension between the two objects, and to obtain conclusive proof of the discovery of a new planet. In less than an hour the brighter star had moved westward about two seconds of time, so that no doubt could be entertained in respect to its nature and position in the solar system. Mr. Hind selected for this planet the name Victoria, with a star and laurel-branch for its symbol. In case, however, this name should be considered objectionable, he proposed that of Clio, which name has been generally preferred by American astronomers.

Remarkable changes of brilliancy in this body have been noticed at the Washington observatory. On the night of November 4, 1850, the planet appeared of the tenth magnitude. On the succeeding night it had diminished to the twelfth magnitude, while the star of comparison exhibited no perceptible change. Differences to a greater amount were observed between the 22d and 25th of February, 1851. On the nights of the 1st and 2d of March, 1851, it appeared as a star of the twelfth magnitude, and was observed without difficulty; the star of comparison being near, and of about the same magnitude. On the night of the 3d, Clio could barely be observed with the faintest illumination, while the same star of comparison used on the nights of the 1st and 2d appeared as before. On the night of the 4th the planet appeared even more brilliant than it did on the nights of the 1st and 2d instant. These changes seem to suggest the probability that the light is reflected with unequal intensity from different sides of this asteroid. Similar differences of magnitude in the other asteroids have been noticed, particularly in Astræa.

The discovery of Victoria was soon followed by that of another asteroid by Dr. Annibal de Gasparis, at the royal observatory, Naples. In this case a star map was not the means of discovering the planet, but its existence was indicated by a series of observations in zones of declination, which had been undertaken for the express purpose of finding new planets. On the 2d of November, 1850, Dr. Gasparis met with the thirteenth asteroid in the constellation Cetus. It was sensibly fainter than stars of the ninth magnitude. M. Le Verrier, to whom was delegated the right of naming this planet, proposed Egeria, the counselor of Numa Pompilius. The orbit is

more inclined to the plane of the ecliptic than that of
most of the other planets.

The next member of the group of small planets in the
order of discovery, was found by Mr. Hind in the constellation Scorpio, on the 19th of May, 1851, and four
days later by Dr. Gasparis, at Naples. It appeared like
a star of between the eighth and ninth magnitudes, with
a full blue light, and seemed to be surrounded by a
faint nebulous envelope or atmosphere, which could not
be perceived about stars of equal brightness. The nature
of this object was satisfactorily established within half
an hour from the first glimpse of it on the 19th of May;
repeated examinations of the vicinity on previous occasions having indicated no star in the position of the
stranger. At the recommendation of Sir John Herschel
the new planet was named Irene, in allusion to the peace
prevailing at that time in Europe; the symbol proposed
being a dove with an olive-branch and star on its head.

On the night of July 29, 1851, another small planet
was discovered by Dr. Gasparis, at Naples, in the course
of his zone observations, commenced with an especial view
to the discovery of new planets. It shone as a fine star
of the ninth magnitude; but, owing to its low situation
in the heavens, was not so generally observed during
its first apparition as some of the other newly-discovered
bodies. Dr. Gasparis named his planet Eunomia, who
in classical mythology was one of the Seasons, a sister of
Irene.

On the 17th of March, 1852, M. de Gasparis, at Naples,

discovered another small planet near the bright star Regulus. It appeared like a small star of the tenth or eleventh magnitude, and has received the name of Psyche. Mr. Hind, of London, narrowly missed the honor of being the first discoverer of this body. On the 29th of January preceding, he entered upon his chart a star of the eleventh magnitude in the place where, according to subsequent computations, this planet ought to have been. The chart was immediately sent to the engraver, and not returned until March 18; but on the evening of that day he discovered that the above star was missing. He immediately commenced a search for the planet, and actually recorded it again on the 20th as a fixed star, but moonlight and unfavorable weather prevented him from establishing its planetary nature before he received the announcement of Dr. Gasparis' discovery.

On the 17th of April, 1852, another planet was discovered near Flora by Mr. R. Luther, at the observatory at Bilk, near Düsseldorf. Professor Argelander, of Bonn, proposed for this planet the name of Thetis, which name was accepted by the discoverer, and has been adopted by astronomers.

On the 25th of June, 1852, Mr. Hind, at London, discovered another planet, having the appearance of a star of the ninth magnitude, and of a yellowish light. Mr. Airy, the astronomer royal, having been requested by Mr. Bishop to select a name for this planet, proposed to call it Melpomene. It is the nearest of the group of asteroids, except Flora, making its revolution in about 1269 days.

On the 22d of August, 1852, Mr. Hind discovered another planet not far from the ecliptic in the constellation Aquarius. It appeared like a star of the ninth magnitude, and exhibited the same yellowish color which was remarked about Melpomene. Mr. Hind having been requested by Mr. Bishop to find a name for this planet, proposed to call it Fortuna.

The next planet was independently discovered by Professor de Gasparis on September 19th, and by M. Chacornac, assistant to M. Valz, at Marseilles, on the 20th of the same month. M. Chacornac was occupied in completing some ecliptic charts of the stars according to a plan adopted by Professor Valz in 1847, and on the night of September 10, he remarked a star of the ninth magnitude in a position where none had been seen before. M. Valz proposed the name Massilia for this object, in which Professor Gasparis, who had a prior claim to the discovery, appears to have concurred. The inclination of its orbit to the ecliptic is less than that of any other known planet, Uranus not excepted.

On the 15th of November, 1852, another planet was discovered, at Paris, by M. Hermann Goldschmidt, an historical painter, residing in that city. M. Arago proposed to call it Lutetia. It resembled a star of the ninth or tenth magnitude.

On the night following the last discovery, November 16th, Mr. Hind, of London, detected a new planet with the assistance of one of the ecliptical star-maps at present in course of publication from Mr. Bishop's observatory. It was not much over the tenth magnitude, which is rather beyond the limit of the Berlin charts. Mr. J. C. Adams, president of the Astronomical Society of London, being requested to name the planet, proposed to call it Calliope.

On the 15th of December, 1852, another planet was detected by Mr. Hind. It had a pale, bluish light, and resembled a star of the tenth or eleventh magnitude, and being not very far from perihelion, is probably one of the faintest members of the group. Mr. Bishop, at the request of Mr. Hind, has selected for this planet the name Thalia.

Thus, within a period of nine months, were discovered eight small planets belonging to the group between Mars and Jupiter, and four of them were discovered by Mr. Hind of London, a fact altogether without precedent in the history of astronomy—a result not of accident, but of a systematic and persevering survey of the heavens.

On the 5th of April, 1853, Professor de Gasparis discovered in the constellation Leo a very minute object, estimated as not brighter than a star of the twelfth magnitude, which on the following evening he recognized as a new planet, in consequence of its proper motion. This discovery was ascribed to the circumstance, that on the 5th of April, 1851, very near to the place where this planet was found, he had observed a star of the twelfth magnitude, which had subsequently vanished; for which

4

reason he was led to examine the neighboring stars with unusual care. Professor Secchi, having been invited by Professor de Gasparis to select a name for this planet, proposed the name of Themis, the same which Professor de Gasparis had originally proposed for Massilia. The mean distance of Themis from the sun is greater than that of any other known asteroid, excepting Hygeia and Euphrosyne, corresponding to a period of 2037 days.

On the night after the preceding discovery, April 6th, M. Chacornac, at Marseilles, discovered another small planet. It appeared of a bluish tint, and of the size of a star of the ninth magnitude. M. Valz proposed to call this planet Phocæa, Marseilles having been founded by a colony from Phocæa.

On the 5th of May, 1853, Mr. R. Luther, director of the observatory at Bilk, near Düsseldorf, discovered a new planet like a star of the eleventh magnitude. This planet was christened by the celebrated Baron von Humboldt, who selected for it the name of Proserpina, with the symbol of a pomegranate and a star in its center.

On the 8th of November, 1853, Mr. Hind discovered another planet within the limits of his ecliptical chart for the third hour of right ascension. It was as bright as stars of the ninth magnitude, and its light appeared remarkably blue. This planet has received the name of Euterpe.

On the 1st of March, 1854, Mr. R. Luther, director of the observatory at Bilk, near Düsseldorf, discovered

another planet. It appeared like a star of the tenth
magnitude, and has received from Professor Encke the
name Bellona; the symbol proposed being a whip and
a lance.

On the same night as the preceding, but about two
hours later, Mr. Albert Marth, at the Regent's Park observatory in London, discovered another planet near
Spica Virginis. It appeared like a star of the tenth or
eleventh magnitude, and Mr. Bishop has proposed for it
the name Amphitrite. On the 2d of March the same
object was independently discovered at the Radcliffe observatory, Oxford, England, by Mr. Pogson, who for
several years has devoted his leisure hours, after the
regular duties of his office are completed, to the formation
of charts of small stars, with the view to the detection of new planets or variable stars. A third independent discovery was made on the 3d of March by M.
Chacornac, assistant observer at the observatory of Paris.
The same impression of the *Times* contained two independent communications from Mr. Hind of London, and
Mr. Johnson of Oxford, each containing the announcement of this discovery. Also, on the 4th of February,
at Marseilles, M. Chacornac noted a star of the tenth
magnitude, which is now wanting in that place, and
which is shown to have been the body first recognized
as a planet by Mr. Marth.

On the 22d of July, 1854, the thirtieth asteroid was
discovered by Mr. Hind at Mr. Bishop's observatory in
Regent's Park, London. It appeared like a star between

the ninth and tenth magnitudes. Professor de Morgan, who was requested by Mr. Bishop to find a name for this planet, has recommended Urania.

On the 1st of September, 1854, the thirty-first asteroid was discovered at the Washington observatory, by Mr. James Ferguson. It was so close to the planet Egeria, of which Mr. Ferguson was in search, that it was observed along with it on the 1st. Another night's observation proved that both were planets, the new one appearing of about the same degree of brightness as Egeria. Mr. Ferguson has been employed for several years with the great equatorial telescope at Washington, and has spent a large portion of his time in observing the places of the newly-discovered asteroids. This is the only instance in which any American astronomer has been the first discoverer of a primary planet. Mr. Bond, of Cambridge, was the first discoverer of the faint satellite of Saturn, and several American astronomers have enjoyed the honor of having first discovered a comet. The honor of naming this new planet was left to Mr. Ferguson, and he has selected the name of Euphrosyne. The period of revolution appears to be greater than that of either of the other asteroids, and its inclination to the ecliptic is greater than any except Pallas.

On the 28th of October, 1854, two new asteroids were discovered at Paris, one of them by M. Goldschmidt, the other by M. Chacornac. The former appeared as a star of somewhat less than the tenth magnitude, and has been named Pomona; the latter somewhat smaller than a star of the ninth magnitude, and has been named Polymnia. The eccentricity of the orbit of Polymnia appears to be greater than that of any other known member of the planetary system, the difference between the distances from the sun at perihelion and aphelion amounting to about the entire diameter of the earth's orbit.

On the evening of April 5th, 1855, M. Chacornac, of the Imperial Observatory of Paris, discovered another small planet of the eleventh magnitude. In about two hours its right ascension had changed nearly five seconds of time, showing clearly its planetary character. On the next day the planet was publicly announced, and was soon found at the other observatories of Europe. This planet has received the name of Circe.

On the 19th of April, 1855, Dr. Luther at Bilk, discovered a new planet of the eleventh magnitude, and on the following day, the notice was communicated by telegraph to the editor of the *Astronomische Nachrichten*. On the 21st it was detected both at the Hamburg and Bonn observatories. At the request of the discoverer, Dr. Peters and M. Rumker gave to the planet the name Leucothea, the protectress of sailors. This planet is one of the most distant of the group of asteroids.

On the 5th of October, 1855, M. Goldschmidt at Paris, discovered a new planet equal in brightness to a star of the eleventh or twelfth magnitude. This new planet has received the name of Atalanta, and the eccentricity of its orbit appears to be greater than that of any other member of the planetary system except Polymnia, and Leda.

On the same evening with the preceding, and nearly at the same hour, the thirty-seventh asteroid was discovered by Dr. Luther at Bilk. It appeared like a star of the tenth magnitude, and on the following day the discovery was communicated by telegraph to the editor of the *Nachrichten*. On the 7th it was seen both at Altona and Hamburg. This planet has received the name of Fides.

On the evening of the 12th of January, 1856, M. Chacornac, at the Paris observatory, discovered a new planet equal in brightness to a star of the ninth or tenth magnitude. This planet has received the name of Leda.

On the 8th of February, 1856, M. Chacornac, of the Paris observatory, discovered a new planet, being the thirty-ninth asteroid, and equal to a star of the eighth or ninth magnitude. This planet has received the name of Lætitia.

On the 31st of March, 1856, a new planet, being the fortieth asteroid, was discovered at Paris by M. Hermann Goldschmidt; and on the 6th of April it was observed both at Altona and Hamburg. It appeared like a star of the ninth or tenth magnitude, and has been named Harmonia.

On reviewing the preceding sketch, we find that Mr. Hind, of London, has been the first discoverer of ten asteroids; M. de Gasparis of seven; Mr. Luther of five; M. Chacornac of five; M. Goldschmidt of four; while Olbers and Hencke, have each discovered two. We also see that, in several instances, the same asteroid has been independently discovered by more than one astronomer.

Among all the astronomers of the present or any former age, Mr. Hind stands pre-eminent for his success in the discovery of new planetary bodies. These discoveries were all made at the private observatory of George Bishop, Esq., which was erected in the year 1836, in Regent's Park, London. The principal instrument of this observatory is an equatorial telescope, constructed by Mr. Dollond, of London, and equipped on the plan known as the English mounting. The solar focus of the telescope is ten feet ten inches, and the clear aperture of the object-glass seven inches. The circles are three feet in diameter; the hour circle reading by verniers to single seconds of time, and the declination circle to ten seconds of arc. This instrument is driven by clock-work; this part of the machinery in particular being very elaborately worked. The telescope is provided with a series of magnifying powers up to 1200.

In the year 1844, Mr. Bishop secured the services of J. R. Hind, Esq., then an assistant in the magnetical department of the Royal Observatory, Greenwich, where he had already distinguished himself by the zeal and ability with which, in addition to his ordinary duties, which were severe, he devoted himself to the labor of observing comets, and calculating the elements of their orbits.

Almost from the time of Mr. Hind's appointment, the observations took that character for which his talents peculiarly fitted him, viz., the search of the heavens for new comets, planets, etc. His labors were almost im-

mediately rewarded with success. Two comets were discovered in 1846, and another in 1847, the latter of which
became visible at noonday, when near its perihelion,
and for which the King of Denmark's gold medal was
awarded.

The search after small planets lying between Mars and
Jupiter was still more successful. His plan for detecting
them was to observe and map all the stars down to the
eleventh magnitude for several degrees on each side of
the ecliptic, and then by a subsequent observation noting
whether any of them seemed to have changed its place,
this being the only planetary characteristic observable.
For the discoveries of Iris and Flora in 1847, a prize on
the Lalande foundation was received from the Academy
of Sciences at Paris in April, 1850; and in February, 1853,
he received the gold medal of the Royal Astronomical
Society of London for his numerous astronomical discoveries, and in particular for his discovery of eight small
planets.

The rapid discovery of thirty-six new asteroids, after a
barren interval of almost forty years from the discovery
of Vesta, is calculated to excite surprise; but it is explained by the diminutive size of the new planets, and
the great increase in the number of observers, as well as
the use of more powerful instruments. Vesta appears
like a star of the sixth magnitude, Pallas of the seventh,
while Ceres and Juno are of the eighth. Of the thirty-
six asteroids more recently discovered, none of them, if
we except perhaps Iris, Flora, and Lætitia, are larger

than the ninth magnitude, while several are as small as the tenth, and three or four scarcely, if ever, rise as high as the tenth magnitude. The reason that Olbers was not more successful in his search was, that he employed a telescope of too feeble power, and did not extend his examination beyond stars of the eighth magnitude.

Some may conclude that the number of asteroids now known is so great, that the discovery of additional ones is a matter of no interest, and is unworthy the attention of astronomers. We regard the question in a very different light. If only one planet had hitherto been discovered between Mars and Jupiter, our idea of the simplicity and perfection of the solar system would have been satisfied, and there might have been found ingenious minds attempting to prove by *à priori* reasoning that no other planets could possibly exist, unless beyond the limits of the orbit of Neptune. But our theory of the solar system, although apparently simple, would not have been the true theory. Every new discovery shows the solar system to be more complex than we had supposed ; and unless we prefer error (provided it has a show of simplicity) to truth, when it appears to our view complex, we shall value every new discovery in the solar system, because it promises to conduct us nearer to the true theory of the universe. Every new asteroid which is discovered, is a new fact to be explained. It presents a new test by which every theory is to be tried. If our theory be false, it is probable that some of these facts may be shown to be inconsistent with it. When the number

4*

of known facts is small, they may all frequently be explained by different and conflicting theories. As the
number of known facts increases, some of them will
probably be found inconsistent with one or the other of
the theories, until at last we reach a fact—the true *experi-
mentum crucis*—which is inconsistent with every theory
but one. Thus the true philosopher, instead of regarding
the rapidly increasing number of asteroids with indiffer-
ence, will watch each new discovery with growing interest, in the hope that it may furnish the key to the true
theory of the solar system.

The following table exhibits a summary of the principal elements of forty asteroids. Column first shows
the number of each planet in the order of its discovery; column second the name of the planet; column
third shows the average distance from the sun (the distance of the earth from the sun being taken as unity) ;
column fourth shows the number of days required to
make one revolution about the sun; column fifth shows
the eccentricity of the orbit, or the quantity by which it
departs from the form of a circle; column sixth shows
the number of degrees by which the plane of the orbit is
inclined to the orbit of the earth ; column seventh shows
the position of the line in which the plane of the orbit
intersects the orbit of the earth; and the last column
shows the position of that point of the planet's orbit
which is nearest the sun.

The existence of forty planets revolving round the sun
at distances closely allied to each other, and differing

ELEMENTS OF THE ASTEROIDS.

No.	NAME.	Distance from sun.	Time of revolution in days.	Eccentricity.	Inclination of orbit.	Long. of ascending node.	Long. of perihelion.
1	Ceres..................	2.766	1680	0.079	11°	81°	150°
2	Pallas.................	2.770	1683	.239	35	173	122
3	Juno	2.668	1592	.256	13	171	54
4	Vesta.................	2.361	1325	.090	7	103	251
5	Astræa	2.577	1511	.189	5	141	136
6	Hebe	2.425	1379	.202	15	139	15
7	Iris...................	2.386	1347	.231	5	260	41
8	Flora	2.201	1193	.157	6	110	33
9	Metis........	2.386	1346	.123	6	68	72
10	Hygeia................	3.149	2041	.101	4	288	227
11	Parthenope	2.448	1399	.098	5	125	317
12	Clio	2.335	1303	.045	8	235	302
13	Egeria	2.577	1512	.085	17	43	120
14	Irene .:...............	2.584	1518	.169	9	187	179
15	Eunomia..............	2.643	1570	.188	12	294	28
16	Psyche................	2.933	1835	.131	3	151	11
17	Thetis	2.484	1430	.131	6	125	259
18	Melpomene...........	2.294	1269	.215	10	150	16
19	Fortuna..............	2.444	1396	.159	2	211	31
20	Massilia..............	2.401	1359	.145	1	207	99
21	Lutetia...............	2.434	1387	.162	3	80	327
22	Calliope..............	2.912	1815	.104	14	67	59
23	Thalia........	2.645	1571	.240	10	68	123
24	Themis................	3.144	2037	.123	1	36	135
25	Phocea	2.401	1359	.253	22	214	303
26	Proserpina...........	2.655	1580	.088	4	46	236
27	Euterpe...............	2.347	1313	.174	2	94	87
28	Bellona...............	2.775	1689	.155	9	145	122
29	Amphitrite............	2.552	1489	.067	6	356	48
30	Urania................	2.366	1329	.126	2	308	31
31	Euphrosyne	3.156	2048	.216	26	31	94
32	Pomona..............	2.583	1516	.083	5	221	195
33	Polymnia.............	2.866	1772	.337	2	9	341
34	Circe	2.673	1596	.114	5	184	158
35	Leucothea............	2.974	1873	.216	8	356	198
36	Atalanta..............	2.757	1672	.300	19	359	43
37	Fides.................	2.654	1580	.180	3	8	66
38	Leda	2.635	1563	.326	6	295	127
39	Lætitia............ ..	2.768	1682	.116	10	157	1
40	Harmonia.............	2.254	1236	.289	5	84	14

from all the other planets in their diminutive size, is one of the most singular phenomena in our solar system.

This fact will appear the more striking if we draw a diagram representing the orbits of all the known planets in their proper proportions. We shall find that while the orbits of Mercury, Venus, the Earth, and Mars are quite detached from each other, and the orbits of Jupiter, Saturn, Uranus, and Neptune are separated by intervals which the imagination can with difficulty grasp, between Mars and Jupiter is a cluster of bodies whose orbits are so interlaced as to suggest the apprehension of frequent and inevitable collision.

The diagram on the following page represents the orbits of nine of these small planetary bodies, designed to be selected so as to afford a tolerable specimen of the whole. The other thirty-one orbits are omitted, to avoid the confusion of so many lines in a single diagram. In one respect this representation is calculated to convey an erroneous impression. All the orbits are represented as situated in the same plane, whereas, in reality, no two of them are situated in the same plane. These planes all pass through the sun, and are inclined to the earth's orbit in angles indicated in column sixth of the preceding Table. One half of each orbit must therefore be *below* the earth's orbit, and the other half *above* it; and in order to indicate as fully as possible the actual position of these orbits, the portion which falls below the plane of the earth's orbit is indicated by a dotted line, while the remainder is indicated by a continuous black line. These orbits, then, do not really intersect each other as represented in the diagram. Indeed, no two of the planetary orbits intersect, although some of them approach within moderate distances of each other.   The orbit of Fortuna approaches the orbit of Metis within less than the Moon's distance from the earth.   The orbit of Massilia approaches almost equally near the orbit of Astræa, and the orbit of Lutetia to that of Juno.

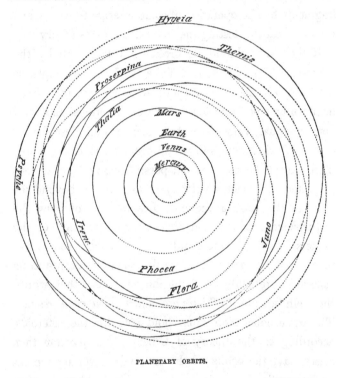

PLANETARY ORBITS.

It is evident, then, at a glance, that these forty small planets sustain to each other a relation different from that of the other members of the solar system.   We see a

family likeness running through the entire group, and it
naturally suggests the idea of a common origin. This
idea, as has been already stated, occurred to the mind of
Olbers after the discovery of the second asteroid, and led
to his celebrated theory that all these bodies originally
constituted a single planet which had been broken into
fragments by the operation of some internal force. Have
we any means of testing the soundness of this theory?

If the earth should be broken into fragments by the
operation of some internal force (such, for example, as
that which causes the eruption of a volcano,) the fragments might be projected in various directions, and with
very unequal velocities; but each would describe an
ellipse of which the sun would occupy one of the foci—
if we except the extreme but possible case of a fragment
projected with such a velocity as to carry it beyond the
limit of the sun's attraction. Leaving out of view the
disturbance arising from the mutual attraction of the
planets, which produces only minute effects, each fragment would continue to describe the same ellipse in its
successive revolutions about the sun; in other words,
these ellipses *would all have a common point of intersection.*
The same conclusion must hold true for the asteroids,
according to the theory of Olbers. The question then
arises, have the orbits of the asteroids a common point
of intersection? A single glance at the above diagram
will settle this question in the negative. But Olbers
replies that the orbits of the planets are disturbed by
their mutual attractions. These orbits should *originally*

have had a common point of intersection, but at each revolution they suffer a slight displacement, until, in the lapse of time, the position of the orbits has become so completely changed as to show scarcely a trace of their original intersection. Is such a result possible? A few simple considerations will satisfy us that if the orbits of the asteroids ever had a common point of intersection, such a result must have belonged to a period of time indefinitely remote.

The line in which the plane of the planet's orbit intersects some other plane selected for common reference is called technically the line of the *nodes*. If the asteroid orbits had ever a common point of intersection, all the nodal lines upon one of the orbits must have coincided. Now, as two of the asteroid orbits are inclined less than one degree to the earth's orbit, we will, for greater convenience, employ the latter as the plane of reference. By referring to our table on page 83, it will be seen that the ascending nodes of the asteroids are distributed, though unequally, through the four quadrants of the circle. Twelve of them lie in the first quadrant, thirteen in the second, eight in the third, and seven in the fourth. The nodes of all the planetary orbits are in constant motion, but the motion for a single year is extremely small. The annual motion of the node of Mercury is ten seconds; that of Venus twenty seconds; Mars twenty-five seconds, etc. The nodes of the asteroids, as far as the computation has been made, move at somewhat similar rates; the most rapid motion known

being about fifty seconds a year. If we suppose the
nodal lines of all these orbits to move steadily toward
each other, it would require in some of them a motion of
fifty seconds a year, continued for more than 6000 years,
to bring them to a coincidence.

It should also be observed, that not only must the
nodes of all the asteroids coincide, but the distance of
the planets from the sun must be the same at that instant.
Now the distances of these planets from the sun, when at
their nodes, differ by more than a hundred millions of
miles; so that to bring them all together requires something more than a change in the position of the nodes.
We may bring about a coincidence, in the case of some
of the asteroids, by supposing the longer diameter of the
elliptic orbit to change its position in the plane of the
orbit.   Such a change does really take place in the case
of every planetary orbit, but with none of the larger
planets does it exceed twenty seconds a year.   This motion for the asteroids, so far as it has been computed, is
somewhat more rapid, amounting, in one instance, to
seventy seconds a year; but even with this motion, it
would require the lapse of five thousand years to bring
about an intersection in the case of many of the asteroid
orbits.   When now it is remembered, that in order to
give a common point of intersection to these forty orbits,
all the nodal lines upon one of the orbits must coincide,
and at the same instant all the distances from the sun
must be equal to each other, we must be prepared to

admit that such an occurrence could only have taken place myriads of years ago.

The preceding difficulties, however, are small in comparison with another which remains to be stated. The orbit of Hygeia completely incloses the orbit of Flora (and indeed several other orbits), and would still inclose them, although the greater diameter of each of them were revolved through an entire circumference, since the least distance of Hygeia from the sun exceeds the greatest distance of Flora. The same is true of Themis, as compared with Flora and several other orbits. The least distance of Hygeia from the sun exceeds the greatest distance of Flora by *more than twenty-five millions* of miles. In order to render an intersection of these orbits possible, we must suppose a great variation of the eccentricity. But the change of the eccentricity of the planetary orbits is exceedingly slow, and the present rate of increase of the eccentricity of Vesta must be continued *twenty-seven thousand years* to render the aphelion distance of that planet equal to the perihelion distance of Hygeia. Moreover, the eccentricity of the orbit of Vesta is now *increasing*, which implies that in past ages the interval between Vesta and Hygeia must have been greater than it is at present; whence the conclusion seems irresistible, that the orbits of Vesta and Hygeia can not have intersected for *several myriads* of years. When the secular variations of the elements of each of the asteroids have been computed, astronomers will be able to assign a limit of time beyond which the intersection of all the asteroid orbits must have occurred, if, indeed, such an intersection ever took place. The discovery of many of these bodies is so recent that, as yet, there has not been sufficient time for such a computation; but, from what we already know, we hazard little in venturing the opinion that, when this computation shall be made, it will appear, that if the asteroid planets ever composed a single body which exploded, as Olbers supposed, such explosion must have occurred *myriads of years ago.* Indeed, the discovery of such a host of asteroids seems to have stripped the theory of Olbers of nearly all the plausibility it possessed when it was originally proposed; and it would seem hardly less reasonable to suppose that the Earth and Venus originally constituted one body, than to admit the same for the forty asteroids.

But if we reject the theory of Olbers, what do we conclude? That the asteroids bear no special relationship to each other? Do they not all clearly indicate a family resemblance? And if so, how do we account for this relationship?

There are several reasons for believing in some peculiar relationship between the asteroids.

1. Unlike the other planets of our system, they are all of diminutive size—the largest of them hardly exceeding one or two hundred miles in diameter. M. Le Verrier, after a close examination of the nature and amount of the influences exerted by the entire group of asteroids upon the nearer planets, Mars and the Earth,

has arrived at the conclusion that the sum total of the matter constituting the small planets situated between the mean distances 2·20 and 3·16 (including undiscovered as well as known asteroids), *can not exceed about one fourth of the mass of the earth.*

2. The asteroids in their position occupy a zone entirely distinct from the other planets of the solar system. Between the orbits of Jupiter and Saturn—between Saturn and Uranus—is an immense interval, furnishing space enough for a host of little bodies to circulate around the sun ; but in not a solitary instance has any such body been found, except between Mars and Jupiter. Some may attempt to account for this circumstance by saying that astronomers have long been watching exclusively this portion of space, and have left all other regions entirely unexplored. An exploration, conducted upon such a principle, is simply a physical impossibility. If there were a small planet between the Earth and Mars, it would have stood the same chance of detection, in the explorations of the past ten years, as if it were situated between Mars and Jupiter ; and, indeed, it would have stood a better chance of detection, inasmuch as it would appear of greater brightness on account of its proximity to us. If there were a small planet circulating between Jupiter and Saturn, it would have stood the same chance of detection as if it had been placed this side of Jupiter, except that it would appear somewhat fainter on account of its increased distance. The fact that we have discovered forty small planets between Mars and Jupiter,

and not a solitary one in any other portion of the solar system, points to something *special* in this region of the heavens. In other words, we have discovered *a limited zone* of little planetary bodies, and have not been able to discover a single body of the same class situated out of this zone.

3. The orbits of these little bodies present some special peculiarities.

If we refer to the table on page 83, we shall perceive that the perihelia of the orbits are not distributed uniformly through the zodiac. In the first quarter of the zodiac we find seventeen perihelia, in the second quarter twelve, in the third quarter six, and in the fourth quarter five. The ascending nodes of the orbits are distributed with greater uniformity. Thus, in the first quadrant we find twelve nodes, in the second thirteen, in the third eight, and in the fourth seven.

The greatest inclination in the case of any of the larger planets is seven degrees; but the inclinations of the orbits of the asteroids range from near zero to thirty-five degrees, the inclination of seventeen of the orbits exceeding seven degrees. The greatest eccentricity in the case of any of the large planets is one fifth; but the eccentricities of the orbits of the asteroids range from near zero to one third.

4. But the most striking peculiarity of these orbits is, that they all lock into one another like the links of a chain, so that if the orbits are supposed to be represented materially as hoops, they all hang together as one system.

The orbits of Hygeia and Themis, being the largest of all the orbits, completely inclose nearly all of them, and lock into but a small number; while the orbits of Massilia, Astræa, Pallas, etc., lock into nearly all of the orbits; so that if we take hold of the orbit of Hygeia (supposed to be a material hoop), it will support the orbits of Iris, Thalia, Calliope, and two or three others, while these in turn lock into and support all the rest. Indeed, if we seize hold of any orbit at random, it will drag all the other orbits along with it. This feature of itself sufficiently distinguishes the asteroid orbits from all the other orbits of the solar system.

If we reject the theory that these asteroids were originally united in one solid body, it seems nevertheless difficult to avoid the conclusion that similar causes have operated in determining the orbits of this zone of planets. It is impossible to assign any cause for these resemblances without adopting some theory respecting the origin of the solar system. The theory of gradual condensation, as developed by Laplace in the nebular hypothesis, affords at least a plausible explanation of these phenomena.

Laplace supposed that the matter composing the bodies of our solar system originally existed in the condition of an immense nebula, extending beyond the limits of the most distant planet—that this nebulous mass had an exceedingly elevated temperature, and a slow rotation on its axis—that the nebula gradually cooled; and as it contracted in dimensions, its velocity of rotation, according

to the principles of mechanics, increased, until the centrifugal force arising from the rotation became equal to
the attraction of the central mass for the exterior zone,
when this zone necessarily became detached from the
central mass. As the central mass continued to contract
in its dimensions, and its velocity of rotation continued to
increase, the centrifugal force again became equal to the
attraction of the central mass for the exterior zone, and a
second zone was detached. Thus, a number of zones of
nebulous matter were successively detached, until, by
condensation, the central mass became of comparatively
small dimensions and great density.

The zones thus successively detached would form concentric rings of vapor, all revolving in the same direction
round the sun. If the particles of each ring continued to
condense without separating from each other, they would
ultimately form a liquid or a solid ring. But generally
each ring of vapor would break up into separate masses,
revolving about the sun with velocities slightly differing
from each other. These masses would assume a spheroidal
form; that this, they would form planets in the state of
vapor. But if one of these masses was large enough to
attract each of the others in succession to itself, the ring
of vapor would be converted into a single spheroidal
mass of vapor, and we should have a single planet of
great mass for each zone of vapor detached. But if no
one of these masses had a preponderating size, they would
all continue to revolve about the sun in independent
orbits, and would form a zone of little planets, such

as we have actually discovered between Mars and Jupiter.

With regard to the actual number of bodies belonging to this zone of planets, we can do little more than conjecture. Already we have one asteroid of the sixth magnitude, one of the seventh, five of the eighth, nineteen of the ninth, and fourteen of the tenth or eleventh. It would require 400 bodies as large as the largest of the asteroids to make a body one fourth of the size of the earth; and, according to Le Verrier, the sum of all the asteroids can not exceed this limit. When we consider the shortness of the period during which stars below the eighth magnitude have been systematically observed, we see room for the discovery of several more planets of the ninth magnitude, and perhaps three or four hundred more of inferior dimensions.

SECTION III.

THE DISCOVERY OF AN EIGHTH SATELLITE OF SATURN.

SATURN has long been known to be attended by seven satellites. If we number these satellites in the order of their distances from the primary, the sixth was discovered by Huygens in 1655, the seventh by Cassini in 1671, the fifth by the same in 1672, the third and fourth also by Cassini in 1684. The first and second were discovered by Dr. Herschel in 1789. The five satellites first discovered may be seen by a ten feet achromatic; but the two interior satellites can only be seen by a very powerful telescope. Sir John Herschel, at the Cape of Good Hope, in 1836 and 1837, repeatedly observed the second satellite with his reflector of 18 inches aperture; and in one instance only, he caught a glimpse of a point of light, which he suspected to be the first satellite. Mr. Lassell, with his large reflector, obtained three observations of the first satellite, in 1846. Sir John Herschel has proposed to distinguish these satellites by proper names, as follows:—

1. Mimas, which makes its revolution in 0d. 22h. 36m.
2. Enceladus,   "   "   1   8   53
3. Tethys,   "   "   1   21   18
4. Dione,   "   "   2   17   44
5. Rhea,   "   "   4   12   25
6. Titan,   "   "   15   22   41
7. Iapetus,   "   "   79   7   54

During the year 1848, this planet was subjected to the most rigid scrutiny, in consequence of the disappearance of its ring, it being presented edgewise to the sun; and on the 16th of September, an eighth satellite was discovered by the Messrs. Bond of Cambridge, Mass. The following is Mr. Bond's account of the discovery. "On the evening of September 16th, we noticed a small star situated nearly in the plane of Saturn's ring, and between the satellites Titan and Iapetus. This circumstance was at that time regarded as accidental; nevertheless, the position of the star, with respect to Saturn, was recorded. The next night favorable for observation was the 18th. While comparing the relative brightness of the satellites, we again noticed the same object, similarly situated with respect to the planet, and we carefully observed its position. But up to this moment, its real nature was scarcely suspected. Measures, carefully made during the evening of the 19th, having proved that the star partook of the retrograde motion of Saturn, we studied that part of the heavens toward which the planet was moving. Each of the stars which it was expected to approach during the two following nights was marked upon a chart, and micrometric measurements fixed its position and distance with respect to neighboring objects.

"The evening of the 20th was cloudy.

"On the 21st the new satellite had approached the planet, and it sensibly changed its position with respect to the stars during the time of observation. Similar observations were repeated on the nights of the 22d and 23d."

5

On the 18th of September, the same object was observed by Mr. Lassell, of Liverpool. The following is Mr. Lassell's account of the discovery. "On the 18th, while I was attentively examining the planet Saturn, I was struck with the appearance of two stars situated on the line of the interior satellites. I supposed that one of them, which I shall designate by c, was the most distant satellite, and the other, which I shall designate by x, a fixed star. In order to prepare myself for the subsequent observations, I made a careful map of its position with respect to some fixed stars in its vicinity, one of which I shall designate by a.

"On the 19th I was surprised to find that the two stars, c and x, already mentioned, had receded from the fixed star a, x remaining exactly on the line of the satellites, but appearing to have approached the planet; while c, although it followed Saturn, had passed to the north of the plane of the orbits of the interior satellites.

"This appearance suggested the idea that x must be a new satellite, and c Iapetus. In order to verify this conjecture, I took the difference of right ascension between x and a, and between c and a. The result showed that in 2h. 36m, x had moved to the west of a by 2s. 47: and that in 1h. 24m, c had moved 1s. 27 to the west of a; which clearly proves that the two stars x and c were in motion. I measured twice at an interval of four hours, the distance of x from the line passing through the interior satellites, and satisfied myself that during this interval, the distance experienced no sensible change. As the motion of Saturn toward the south was 18″ in 4 hours, it would plainly have left the point x behind, if it had been a fixed star. The conclusion is inevitable; x is a satellite hitherto undiscovered. This is explained by its being a very faint object, even in my telescope of 24 inches aperture; and it may experience variations of light, which render it invisible in some parts of its orbit.

"I obtained two other observations of the satellite on the 21st and 22d. In the former case, the elongation to the east of the planet was 3′ 54″; and in the latter 3′ 27″; the star followed sensibly the line of the interior satellites."

Thus it appears that a most important discovery was made independently and almost simultaneously on opposite sides of the Atlantic—but Mr. Bond has an unequivocal priority of *two days*. This is the first addition to our planetary system made by an American astronomer. The orbit of the new satellite lies between Titan and Iapetus, and serves to fill up a large chasm before existing. It is fainter than either of the two interior satellites discovered by Sir William Herschel. Its time of revolution is about 21d. 4h. 20m.; its semi-axis at the mean distance of Saturn 214″, indicating a real distance of about 940,000 miles, and Messrs. Bond and Lassell have concurred in giving it the name of Hyperion.

It will be observed that there is still a large gap between Hyperion and Iapetus, rendering it not improbable that other satellites yet remain to be discovered.

SECTION IV.

ON THE SATELLITES OF URANUS.

URANUS was discovered to be a planet by Sir William Herschel in 1781, and in 1787 he discovered two satellites, whose periods were satisfactorily determined by his subsequent observations. In 1797 he announced the discovery of four additional satellites, viz., one within the orbits of both the former two; one intermediate between the two; and two exterior to both of them, but the periods of these satellites he acknowledged to be very uncertain. In his last paper on this subject, communicated to the Royal Society in 1815, he says, "that there are additional satellites, besides the two principal larger ones, I can have no doubt; but to determine their number and situation, will probably require an increase of illuminating power in our telescopes."

In 1834, Sir John Herschel published a paper containing a thorough discussion of his father's observations, together with his own, upon the two satellites first discovered; and he adds, "of other satellites than these two, I have no evidence."

In the year 1838, Dr. Lamont, of Munich, published a few observations of the two brighter satellites of Uranus, and states that he had seen only one additional satellite,

and that but in a single instance.   This satellite he considered to be the most remote of the six enumerated by
Herschel.

With the exception, therefore, of the solitary observation of Dr. Lamont, the only evidence we have had (until
recently) of the existence of more than two satellites
of Uranus was derived from the observations of
Sir William Herschel; and he would not pronounce a
decided opinion as to their number or their periods of
revolution.

At last, in the autumn of 1847, Mr. Lassell, of Liverpool, and M. Struve, at Pulkova, obtained unequivocal
evidence of the existence of a third satellite. The orbit
of this satellite was evidently smaller than that of either of
the two bright ones; yet the period indicated by Lassell's
observations did not agree with that deduced by Struve ;
and both differed from the interior satellite of Sir
William Herschel.   While Lassell's observations indicated a period of about two days, Struve deduced
from his observations a period of four days; and the
time assigned by Herschel to his interior satellite was
nearly six days.   Thus the question seemed involved in
total confusion, and the honest enquirer might well be
puzzled to decide whether there existed three satellites,
or only one, interior to the two brighter ones.

In the autumn of 1851, Mr. Lassell succeeded in
settling this vexed question.   On ten different nights in
the months of October, November and December, he saw
simultaneously four satellites, and recorded their positions.

The intervals were so short as to enable him to identify each satellite without danger of mistake. These satellites were the two brighter ones discovered by Herschel, and two interior ones, whose periods are about two and four days respectively.

In the autumn of 1852, Mr. Lassell transported his telescope to the island of Malta, in the Mediterranean, for the purpose of securing the advantage of a lower latitude and a purer sky, and here he succeeded in obtaining a very complete series of observations of the satellites of Uranus. The nearest satellite was observed on 24 different nights; the second satellite on 23 nights; and the two brighter satellites on 26 nights. By combining these observations with those of 1847 and 1851, we are able to assign the dimensions of their orbits and their periods of revolution with great precision. The period of the nearest satellite is 2.520345 days; or 2d. 12h. 29m. 20s. 66. The period of the second satellite is 4.144537 days; or 4d. 3h. 28m. 8s. 00.

From the combination of all the observations between 1787 and 1848, Mr. Adams has determined the period of the third satellite to be 8d. 16h. 56m. 24s. 88; and in the same manner he has determined the period of the fourth satellite to be 13d. 11h. 6m. 55s. 21.

Let us now examine the observations of Sir William Herschel, and see what light is shed upon them by the information recently obtained. There are four instances in which he observed what he called "an interior satellite."

On the 18th of January, 1790, he recorded an object "excessively faint, and only seen by glimpses—two diameters of the planet following."

On the 27th of March, 1794, he recorded as follows: "I had many glimpses of small stars or supposed satellites, but not one of them could I see for any constancy. They were only lucid glimpses."

On the 15th of February, 1798, an interior satellite was seen about its greatest northern elongation, between the planet and the second satellite. On the 16th the place where the satellite had been the day before was scrupulously examined, and as there was no star remaining in that place, the removal of the interior satellite from its former position was ascertained.

On the 17th of April, 1801, the interior satellite was seen about its greatest southern elongation. On the following night, no star was visible in the same position.

It must be considered doubtful whether in 1790 and 1794, Sir William Herschel really saw an interior satellite. The observations of 1798 and 1801 appear to be more reliable; but the interval between these observations does not correspond very well with the motion of either the first or second satellite of Lassell. It is possible that one of these observations may have been made upon the first satellite, and the other upon the second.

In one instance Sir William Herschel saw what he called an intermediate satellite.

On the 26th of March, 1794, a supposed satellite was

seen more than two hours. On the following evening
it was gone from the place where he had seen it the
night previous.

There are five instances in which Herschel observed
what he called "an exterior satellite."

On the 9th of February, 1790, a supposed third satellite was remarked in a line with the planet and the
second satellite, and about twice the distance of the
second satellite. On the 12th the supposed satellite of
the 9th was not in the place where it was then seen.

On the 31st of January, 1791, a satellite was observed
in opposition to the second, and at about double the distance from the planet.

On the 26th of February, 1792, a star toward the south,
at double the distance of the first satellite, was pointed
out, but it has not been accounted for in succeeding
observations.

On the 5th of March, 1796, Herschel recorded, " I suspected a very small star between c and h, which was not
there last night.   I had a pretty certain glimpse of it.
With 600 I see the satellite c better than before, but can
not perceive the suspected small star."

On the 11th of February, 1798, an exterior satellite
was observed, and its situation measured.

There are four instances in which Herschel observed
what he called " the most distant satellite."

On the 28th of February, 1794, Herschel remarked a
very small star which he did not see on the 26th.

On the 27th of March, 1794, he obtained lucid

glimpses of some stars south, but not one of them could he see for any constancy.

On the 28th of March, 1797, he observed an excessively small star about four times the distance of the second satellite, which he did not remember to have seen on the 25th.

On the 16th of February, 1798, a very faint satellite was observed, and from its distance was supposed to be a little before or after its greatest southern elongation. It was so faint that a small alteration in the clearness of the air rendered it invisible. On the 18th this satellite was seen again, and being nearer the planet than it was on the 16th, it was supposed to be on its return from the greatest southern elongation. It was also ascertained on the 18th that it had left the place where it was seen on the 16th.

On the strength of the preceding observations it has been generally admitted that Uranus has six satellites; but Mr. Lassell, during his residence at Malta, carefully scrutinized the neighborhood of the planet to the distance of five minutes from his center, for the discovery of satellites. In the course of this scrutiny, he made many measurements and diagrams of the positions of small points of light, which all turned out to be stars, and he adds, "I can not now resist the conviction, amounting indeed in my own mind to certainty, that Uranus has no other satellites (except four) visible with my eye and optical means. In other words, I am fully persuaded that either he has no other satellites than these four, or if he has, they remain yet to be discovered."

The following diagram represents the apparent orbits
of the four satellites of Uranus for November, 1852.
The major axes of all the orbits are nearly perpendicular
to the ecliptic. The minor axis of each of the apparent
orbits is about three quarters of the major axis, but the
true orbits of the satellites differ very little from circles.

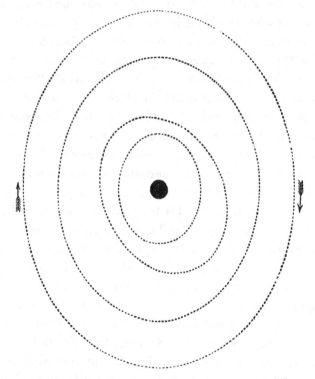

APPARENT ORBITS OF THE SATELLITES OF URANUS FOR NOVEMBER, 1852.

The planet Uranus is fast approaching the position
most favorable for the observation of its satellites. It has

now a north declination of 17 degrees, which will continue to increase until 1868; and in 1862, the orbits of the satellites will appear sensibly circular. It is evident then that the period is near at hand when this planet can be observed more advantageously than at any former time since the first observations of Sir William Herschel, and it is presumed that astronomers will not fail to improve this opportunity to attain all the information within the reach of our instruments, both respecting the physical peculiarities of the primary, and the number and period of its satellites.

SECTION V.

DISCOVERY OF A NEW RING TO SATURN.

THE planet Saturn has long been known to be surrounded by two rings. In the year 1610, Galileo, with a very indifferent telescope, saw the planet like a central globe between two smaller ones. In 1656, Huygens gave the true explanation of these appearances, and showed that the planet is surrounded by a luminous ring, in the center of which it is suspended. In 1675, Cassini saw the dark elliptic line which divides the ring into two concentric portions, and he noted the unequal brilliancy of the rings, the inner one being the brightest. Several astronomers have suspected the existence of more than two rings about Saturn. On the 25th of April, 1837, Professor Encke, at Berlin, saw the outer ring divided into two nearly equal parts, and several divisions were recognized on the inner edge of the inner ring. On the 7th of September, 1843, Messrs. Lassell and Dawes saw what they considered to be a division of the outer ring. The division of the outer ring has also been noticed by Captain Kater in England, M. De Vico, at Rome, and Quetelet, at Paris. The newly-discovered ring of Saturn can not be classed with the subdivisions of the old ring, as it lies within its inner edge.

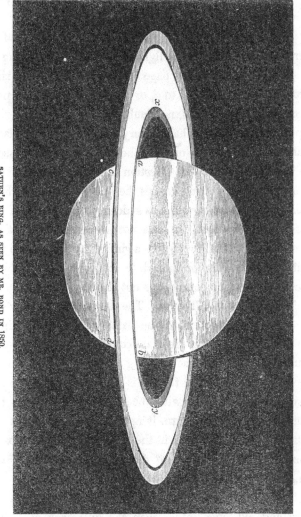

SATURN'S RING, AS SEEN BY MR. BOND IN 1850.

The following is the account of the grand discovery of
a new ring by the Messrs. Bond, at Cambridge:

"November 11, 1850.   We notice to-night with full
certainty the filling up of light inside of the inner ring at
x and *y* (see drawing on page 109).   Also where the ring
crosses the ball from *c* to *d*, or apparently below its projection, is a dark band, no doubt the shadow of the ring
upon the ball; but what is very singular, there is also a
dark line from *a* to *b*, or *above* the ring, very plainly seen,
so that there can be no question as to the line where the
upper edge of the ring crosses the ball.   The light which
fills the space at *x* and *y*, is suddenly terminated on the
side toward the ball.   It does not arise from any optical
deception, for this would give a similar appearance to the
outside of the ring, or indeed to the edge of any object
we look at, which certainly is not the case.

"November 15, 7h 50m.   The new ring is sharply defined on the edge next to the ball.   W. C. Bond thinks
he sees the new ring clear of connection with the old.
But the side next to the old ring is not so definite as that
next to the planet, so that it is not certain whether the
new ring is connected with the old or not.   Where the
dusky ring crosses Saturn, it appears a little wider at the
outside of the ball than in the center.   Where it crosses
the ball, it is not quite so dark as the shadow of the
ring.

"8h P. M. Can not be sure of a division between the
new and old rings (other than the difference of light).
Once or twice with the higher powers, one was suspected.

On further examination we agree that the dark ring is narrower than the outer ring. Its inner edge may be as far from the inner edge of the broad ring as two thirds of the breadth of the outer ring."

The appearances above described had been noticed on many occasions prior to the above dates, but their true explanation was first ascertained on the evening of November 15th. The fact of the existence of a dusky ring hitherto unknown contained in the space between the old ring and the ball could no longer be questioned. Observations continued to the 7th of January, 1851, fully confirmed the deductions which the Messrs. Bond had drawn from those of the 11th and 15th of November.

Similar appearances were noticed by the Rev. W. R. Dawes at his observatory near Maidstone in England, on the 29th of November, 1850. His telescope was a refractor by Merz & Son of Munich, having an aperture of six inches. He states that on the 29th he observed a shading like twilight at the inner portions of the inner ring. Also an exceedingly narrow black line on the ball at the southern edge of the ring where it crosses the planet, and it was slightly broader at the east and west edges of the ball than near its middle. On the 3d of December he again saw the same phenomenon, as did also Mr. Lassell of Liverpool, who was on a visit to Mr. Dawes's observatory. Mr. Lassell stated that the space between the inner diameter of the inner ring and the ball, was half covered by an appearance as if a crape vail had been thrown over it. There is a dark inner edge to the bright ring, which shades off a little. The dark ring is there of a uniform gray color; the sky is seen black between Bond's ring and the planet.

An appearance somewhat similar to that described by Mr. Bond was seen by Dr. Galle at the Berlin observatory in the years 1838 and 1839. The following are the observations of Dr. Galle:

"1838, May 25th. The dark space between Saturn and his ring seemed to consist, as far as its middle, of the gradual extension of the inner edge of the ring into the darkness, so that the fading of this inner ring has considerable breadth.

"June 10. The inner edges of the first ring fade away gradually into the dark interval between the ring and the ball. It seemed that the ring, from the beginning of the shading inclusive, extends over nearly half the space toward the ball of Saturn.

"June 15. The fading of the inner ring toward Saturn, as on June 10."

The memoir containing these observations was published in the Transactions of the Berlin Academy of Sciences for 1838; it attracted, however, but little attention until after the announcement of the discovery of the Messrs. Bond.

Professor A. Secchi, of the observatory at Rome, remarks that this species of penumbra in the interior of the ring of Saturn existed in the year 1828, at which time it was seen with the telescope of Cauchoix, then recently received. The dark ring was perceived, although its nature was not comprehended.

The preceding observations naturally suggest the inquiry, whether this newly-discovered ring is one of recent formation, or has it existed from time immemorial? Sir William Herschel diligently observed this planet; yet only on four occasions did he notice any thing remarkable in the interior portion of the ring beyond a slight shading off, which is represented in one of his figures of the planet, but is not thought worthy of particular description.

Professor Struve remarks (Memoirs of the Astronomical Society, vol. 2), " the inner ring toward the planet seems less distinctly limited, and to grow fainter, so that I am inclined to think that the inner edge is less regular than the others." Neither did Sir J. Herschel at the Cape of Good Hope notice any thing like the present appearances, though Saturn was nearly in the zenith, and the climate so favorable.

In a paper published in the Memoirs of the Academy of Sciences of St. Petersburg, M. Otto Struve attempts to show that the observations of various astronomers anterior to the year 1838, afford unequivocal indications of the existence of the obscure ring. He remarks that the early astronomers generally make mention of a dark belt, (termed the equatorial belt) which was seen passing across the body of the planet, *immediately contiguous* to the inner edge of the interior bright ring. J. Cassini showed in 1715 that from its slight curvature the equa-

torial belt could not be situate close to the surface of the
planet. In 1723, Halley remarked that the dusky line
which in 1720 he observed to accompany the inner edge
of the ring across the disc, continued close to the same,
though the breadth of the ellipse had considerably increased since that time. M. Struve remarks that the
projection of the obscure ring upon the body of the planet
agrees with the position assigned by the earlier observers
to the equatorial belt, and that no trace of the latter is
visible upon the planet in the present day. He therefore
arrives at the conclusion that the obscure ring has existed around the planet, at least since the time when
telescopes were first constructed of sufficient optical
power to exhibit the belts upon his disc.

The breadth of the space which separates the inner edge
of the interior bright ring and the body of the planet having exhibited a considerable discordance as measured by
himself in 1851, and by his father in 1826, M. Otto Struve
instituted a careful examination of the various measurements of preceding astronomers relative to the dimensions
of the planet and the rings. By a comparison of the micrometrical measures of Huygens, Cassini, Bradley, Herschel, W. Struve, Encke and Galle, with the corresponding measures executed by himself, he found *that the
inner edge of the interior bright ring is gradually approaching the body of the planet, while at the same time
the total breadth of the two bright rings is constantly
increasing.* The earlier measures employed in this comparison are, however, founded upon certain assumptions

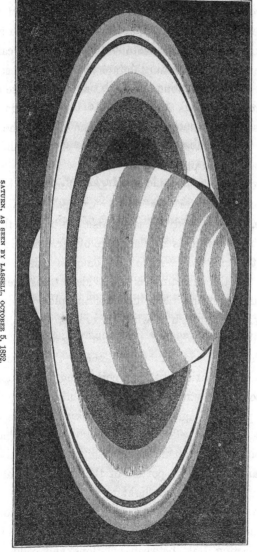

SATURN, AS SEEN BY LASSELL, OCTOBER 5, 1852.

relative to the effects of irradiation, and can not therefore be regarded as altogether trustworthy. From a comparison similar to that instituted in the previous case, M.
Struve infers, that during the interval which elapsed between the observations of J. D. Cassini and those of Sir
William Herschel, the breadth of the inner ring had
increased in a more rapid ratio than that of the outer
ring. The subsequent measures seem to indicate a
reversal of this process.

By observations made by Mr. Lassell at Malta in 1852,
it appears that the new ring is transparent to such a
degree, that the body of the planet can be seen through
it. The following is the language of Mr. Lassell: "Perhaps the most remarkable phenomenon which I now
notice for the first time is the evident transparency of
the obscure ring; both limbs of the planet being distinctly seen through it where it crosses the ball, quite
through to the edge of the inner bright ring. To my
apprehension I can not better describe the entire aspect
of the obscure ring than by comparing it to an annulus
of black crape stretched within the bright ring, which,
when projected against the black sky, as at the curve,
would, from its reflecting some light, appear of a dark
gray shade; and when projected on the ball, would, from
the transmission of a portion of the reflected light of the
ball, appear of a much lighter gray. What the precise
nature of this marvelous appendage can be would be
an interesting subject of speculation, exhibiting as it were
a connecting link between nebulous and solid matter.

The precise definition of its edges renders it unlike any other specimen of nebulæ; while on the other hand its certain translucency deprives it of all resemblance to the other solid bodies of our system." The drawing on page 115 is copied from one furnished by Mr. Lassell from his observations made at Malta in 1852.

Mr. G. P. Bond maintains that Saturn's ring is in a fluid state, or at least does not cohere strongly. The following are some of the considerations upon which he founds this conclusion. Several observers, among whom are Kater, Encke, De Vico, and Lassell, have seen divisions both of the outer and inner ring. On the other hand, some of the best telescopes in the world, in the hands of Struve, Bessel, Sir John Herschel, and others, have given no indication of more than one division, when the planet has appeared under the most perfect definition. The fact also that the divisions on both rings have not usually been visible together, and that the telescopes which have shown distinctly several intervals in the old ring, have failed to reveal the new inner ring, while the latter is now seen, but not the former, indicates some real alteration in the disposition of the material of the rings.

These facts are most easily explained by supposing that the rings are in a fluid state, and within certain limits change their form and position in obedience to the laws of equilibrium of rotating bodies. This hypothesis is favored by considerations drawn from the state of the forces acting on the rings. On the assumption that the matter, of which the ring is composed, is in a solid state, we may compute for any point on its surface the sum of the attractions of the whole ring and of Saturn. The centrifugal force generated by its rotation, may thus be determined from the condition that the particle must remain on the surface. Now, in the case of a solid ring, particles on the inner and outer edges must have the same period of rotation. This condition limits the breadth of the ring; for, if it be found necessary for the inner and outer edges to have different times of rotation, this can be accomplished only by a division of the ring into two or more parts. From careful computations, he has inferred the necessity of admitting a large number of rings, provided they are solid. But there are numerous objections to admitting a large number of small rings near each other, and to avoid these difficulties, he adopts the hypothesis of a fluid ring. If in its normal condition the ring has but one division, as is commonly seen, under peculiar circumstances it might be anticipated that the preservation of their equilibrium would require a separation in some regions of either the inner or outer ring; this would explain the fact of occasional sub-divisions being seen. Their being visible for but a short time, and then disappearing, to the most powerful tele-scope, is accounted for by the removal of the sources of disturbance when the parts thrown off would re-unite.

Professor Peirce has undertaken to show, from purely mechanical considerations, that Saturn's ring can not be

solid. He maintains unconditionally that there is no con-
ceivable form of irregularity and no combination of irregularities consistent with an actual ring, which would
serve to retain it permanently about the primary, if it
were solid.   He maintains that Laplace's statement of the
sustaining power of an irregularity, was a careless suggestion, which was dropped at random, and never subjected to the scrutiny of a rigid analysis.   Moreover the
fluid ring can not be regarded as one of real permanence
without the aid of foreign support.   This support he
finds in the action of the satellites.   The satellites are
constantly disturbing the ring, and yet they sustain it in
the very act of perturbation.   No planet, he thinks, can
have a ring unless it is surrounded by a sufficient number
of properly arranged satellites.

We might hope to obtain important information respecting the constitution of the rings of Saturn, if it could
be observed in its passage over some bright star.   In that
case the light of the star might be seen through each successive opening between ring and ring, provided the
width of the opening were sufficient to allow the visual
ray to clear the thickness of the rings.   It would be important to notice whether the light of the star disappeared
suddenly behind each of the rings in succession, or
whether there was any appearance of refraction.   Whis-
ton informs us that Dr. Clarke's father saw a fixed star
between the ring and the body of the planet; and Professor Robison mentions a star's having been seen in the

interval between the two bright rings.*   Opportunities
of this kind are of exceedingly rare occurrence, and it is
of the highest importance that they should always be improved by nice micrometrical measurements.

* Smyth's Celestial Cycle, v. i, page 193.

CHAPTER II.

SECTION I.

THE GREAT COMET OF 1843.

MODERN astronomers were generally agreed that the ancient accounts of comets were greatly exaggerated; for, said they, since we have had careful and scientific observers, the appalling comets of antiquity have disappeared. What then shall we say of a comet in the nineteenth century, rivaling the noonday splendor of the sun?

On Tuesday, the 28th of February, 1843, a brilliant body resembling a comet, situated near the sun, was seen in broad daylight, by numerous observers in various parts of the world. It was seen in each of the New England States (except, perhaps, Rhode Island), in Delaware, at Halifax, N. S., in Mexico, in Italy, and it is said also in the East and West Indies. It was seen in New England as early as half-past seven in the morning, and continued till after 3 P.M., when the sky became considerably obscured by clouds and haziness. The appearance was that of a luminous globular body with a short train—the whole taken together being found by measurement about one degree in length. The head of the comet, as observed by the naked eye, appeared circular; its light equal to that of the moon at midnight in a clear sky; and its apparent size about one eighth the area of the full moon. Some of the observers compared it to a small cloud strongly illuminated by the sun. The train was of a paler light, gradually diverging from the nucleus, and melting away into the brilliant sky. An observer at Woodstock, Vt., viewed it through a common three feet telescope. It presented a distinct and most beautiful appearance—exhibiting a very white and bright nucleus, and a tail dividing near the nucleus into two separate branches. At Portland, Me., Captain Clark measured the distance of the nucleus from the sun, the only measurement (with one exception) known to have been made in any part of the globe before the 3d of March. At 3h. 2m. 15s., mean time, the distance of the sun's furthest limb from the nearest limb of the nucleus was 4° 6' 15".

In Mexico, Lat. 26° 8' N., Long. 106° 48' W. of Greenwich, the comet was observed from nine in the morning until sunset, by Mr. Bowring, and the altitude of the comet repeatedly measured with a sextant. Professor Bianchi, of Modena, in Italy, writes that on the 28th of February, the comet was seen by numbers at Boulogne, Parma and Genoa.

It is said also that Captain Ray, being at Conception, in South America, saw the comet on the 27th of February, east of the sun, distant about one sixth of his diameter.

The comet was seen at Pernambuco, Brazil, and in Van Dieman's Land, on the 1st of March. On the second, it was seen in great brilliancy at St. Thomas, and by various navigators in the equatorial regions. On the evening of the 3d, it was noticed at Key West, and excited much attention. On the 4th, it was seen by a few in the latitude of New York, and on Sunday evening the 5th, it was noticed very generally. From this date, until about the close of the month, it presented a most magnificent spectacle every clear evening in the absence of the moon. As seen near the equator, the tail had a darkish line from its head through the center to the end. It was occasionally brilliant enough to throw a strong light upon the sea. The greatest length of tail was, about the 5th of March, 69° as measured with a sextant, and it was observed to have considerable curvature. One observer describes it as an elongated birch rod slightly curved, and having a breadth of one degree. At the Cape of Good Hope, on the 3d, it is described as a double tail, about 25° in length, the two streamers making with each other an angle of about a quarter of a degree, and proceeding from the head in perfectly straight lines. The greatest length of tail observed in the United States, was about 50°. The curvature of the tail upward, though very noticeable, scarcely exceeded two degrees. The first observation of the nucleus (with the exception of the noonday observations) is believed to have been made at the

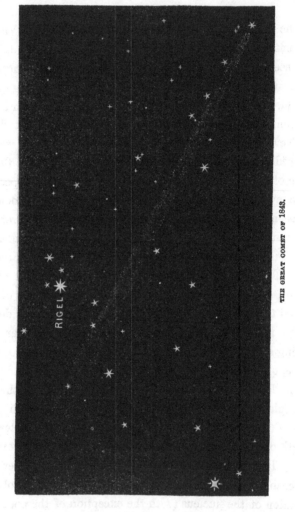

THE GREAT COMET OF 1843.

Cape of Good Hope, on the 3d, after which it was observed regularly until its disappearance. At Trevandrun,
in India, it was observed from the 6th; at Cambridge,
Mass., it was observed on the 9th, and at numerous places
on the 11th. The first European observation of the nucleus was made on the 17th at Rome and Naples, after
which it was seen at most of the continental observatories
till the end of March.

The comet nowhere continued visible many days. It
was seldom seen in Europe after the 1st of April. The
last observation at Naples was on the 7th. Only one
later observation has been announced from Europe. This
was at Berlin, on the 15th, when Professor Encke thought
he caught a faint glimpse of the comet, but it could not
be found again on the subsequent evening. The most
complete series of observations in this country was made
by Messrs. Walker and Kendall of Philadelphia, where
the comet was followed until April 10th. At Hudson,
Ohio, the comet was seen for the last time April 6th.

A great many astronomers have computed the orbit of
this comet, and have obtained very extraordinary results.
The comet receded from the sun almost in a straight line,
so that it required careful observations to determine in
which direction the comet passed round the sun, and some
at first obtained a direct orbit, when it should have been
retrograde. The perihelion distance was extremely small,
very little exceeding the sun's radius. Some have obtained a smaller quantity than this, but such a supposition seems to involve an impossibility. It is certain,

however, that the comet almost *grazed* the sun; perhaps some portion of its nebulosity may have come into direct collision with it. The best orbits give a perihelion distance of ·0056, or a distance of 90,000 miles from the sun's surface, which is equal to about one fifth of the sun's radius.

The velocity with which the comet whirled round the sun at the instant of perihelion was prodigious. This was such as, if continued, would have carried it round the sun in two hours and a half; in fact, it did go more than half round the sun in this time. In one day (that is from twelve hours before, to twelve hours after perihelion passage), it made 291 degrees of anomaly; in other words, it made more than three quarters of its circuit round the sun. In 40 days, the period of its visibility, it had described 173 degrees from perihelion; while to describe the next seven degrees requires a period of many years, and perhaps centuries.

The head of this comet was exceedingly small in comparison with its tail. When first discovered, many were unwilling to believe it a comet, because it had no head. The head was probably nowhere seen by the naked eye after the first days of March. At the close of March the head was so faint as to render observations somewhat difficult even with a good telescope, while the tail might still be followed by the naked eye about thirty degrees. Bessel remarked that "this comet seemed to have exhausted its head in the manufacture of its tail." It is not, however, to be hence inferred, that the tail was really brighter than the head, only more conspicuous from its greater size. A large object, though faint, is much more noticeable than a small one of intenser light. Thus the milky-way is little more than an assemblage of faint stars, no one of which singly would make any distinct impression upon the naked eye.

The nearest approach of the comet's head to the earth was about 80 millions of miles, being very soon after the perihelion passage. The absolute diameter of the neb-ulosity surrounding the head was about 36,000 miles. The apparent length of the tail has already been stated; but the absolute dimensions were still more extraordinary. The following table was computed from the best ob-servations, and shows the progress of the formation of the tail after perihelion. The comet passed its perihelion on the afternoon of February the 27th, Greenwich time.

	Length of tail. Miles.			Length of tail. Miles.
February 28,	35,000,000		March 15,	105,000,000
March 1,	55,000,000		" 17,	106,000,000
" 2,	70,000,000		" 18,	106,000,000
" 3,	82,000,000		" 19,	107,000,000
" 4,	91,000,000		" 20,	107,000,000
" 5,	96,000,000		" 21,	108,000,000
" 6,	99,000,000		" 24,	108,000,000
" 7,	100,000,000		" 26,	106,000,000
" 8,	101,000,000		" 27,	105,000,000
" 9,	102,000,000		" 30,	101,000,000
" 11,	103,000,000		" 31,	98,000,000
" 13,	104,000,000		April 1,	95,000,000

The visible portion of the tail attained its greatest length early in March, remained nearly stationary for some time, and during the first week of April suddenly

disappeared, from increased distance, without any great
diminution of length.   The comet doubtless had a tail
before perihelion, but it seems physically impossible that
this should have formed any part of that which was seen
after perihelion.   The tail was turned nearly toward the
earth on the night of February 27th, in such a direction,
that had it reached the earth's orbit, it would have passed
fifteen millions of miles south of us.   Its length was,
however, at that time, insufficient to reach any considerable part of the distance to the earth.

What gave the comet its extraordinary brilliancy on
the 28th of February?   Evidently its proximity to the
sun.   The day before it had almost *grazed* the sun's
disc.   The heat it received, according to the computations of Sir John Herschel, must have been 47,000 times
that received by the earth from a vertical sun.   The
rays of the sun united in the focus of a lens thirty two
inches in diameter, and six feet eight inches focal length,
have melted carnelian, agate and rock crystal.   The heat
to which the comet was subjected must have exceeded by
twenty-five times that in the focus of such a lens.   Such
a temperature would have converted into vapor almost
every substance on the earth's surface ; and if any thing
retained the solid form, it would be in a state of intense
ignition.   The comet on the 28th of February was *red
hot*—and for some days after its perihelion, it retained a
peculiar *fiery appearance*.   In the equatorial regions, the
tail is described as resembling " a stream of fire from a
furnace."   The reason why the tail was seen on the 28th,

was, that it was turned almost directly toward the earth, and we therefore saw it in the direction of *its length*.

Has this comet ever been seen before? In the year 1668 a comet appeared very similar to the present one. Cassini saw the tail at Boulogne, March 10, and subsequently an hour after sunset; but the head was plunged in the rays of the sun. The tail was 45° in length, and extended almost horizontally from west to south. It was so brilliant that its image was distinctly seen reflected from the sea; but this brightness lasted only three days. The head was small and faint, and difficult to be seen. Professor Henderson, of Edinburg, computed the orbit of this comet, and obtained a result very similar to the early results with the comet of 1843. He then assumed the elements of the latter comet, and from them computed the places in 1668. The results accorded pretty well with the observations. Mr. Peterson has made the comparison with corrected elements, and found the accordance still better. On the whole, it appears that the comets of 1688 and 1843 pursued nearly the *same path*, and exhibited nearly the *same appearances*.

A similar comet was seen in 1689. It was not seen in Europe, but was observed at Pekin, and was most brilliant in the southern hemisphere, where the tail attained the length of about 68°. The head was bright, but the tail was paler, and had the shape of a huge saber curved at the extremity. The orbit of this comet was computed by Pingré, and its elements agree pretty well with those

of the present comet, except the inclination, which was
too great. Professor Peirce, of Cambridge, has re-com-
puted the orbit and obtained a much better coincidence.
In short, it appears that the three comets of 1668, 1689,
and 1843, all pursued nearly the same path, and presented
somewhat similar appearances. What are we to infer?
Mr. Walker inferred that these were different appearances
of the same comet, with a period of $21\frac{7}{8}$ years, the comet
having made seven revolutions from 1689 to 1843. But
why was it not seen at either of the intermediate returns?
Because (it is said) the position of the comet was unfavorable for observation. From the position of its orbit, the
comet will always have considerable southern declination,
which is unfavorable to its being seen in the northern
hemisphere. This answer is not entirely satisfactory, as
it is improbable that so prodigious a tail should pass un-
noticed during six successive returns.

Professor Schumacher is inclined to regard this comet
as identical only with that of 1668, giving it a period of
175 years. Professor Peirce has made a comparison of
all the best observations, and his result is that an ellipse
of short period is inadmissible. The observations may
all be represented by a parabola within the usual limits
of the errors of cometary observations. An ellipse of
about 180 years represents them a little better. Professor Hubbard, of the Washington Observatory, has recently undertaken a thorough discussion of all the observations of this comet, and has computed the effect due
to the perturbations of the planets. He finds the most

probable orbit to be an ellipse of over 500 years. We seem obliged then, entirely to reject any short period, as 20 or 30 years, and there seems but slight probability that this comet is identical with that of 1668. It is doubtful whether the period of this comet can be certainly determined until it has been again observed on its return to the sun.

The following circumstances invest the comet of 1843 with peculiar interest.

1. Its small perihelion distance; being as small as that of any comet whose orbit has been computed, and nearly as small as is physically possible.

And 2. Its prodigious length of tail; being equal to that of any comet hitherto observed.

SECTION II.

On the 22d of November, 1843, a telescopic comet was discovered by M. Faye, of the Paris observatory. It had a brilliant nucleus, and a short tail like a fan about four minutes in length. It was re-discovered in this country on the 27th of December, by Mr. J. S. Hubbard, at New Haven. The comet was closely watched at the European observatories during December and January, and at the Pulkova observatory it was followed until the 10th of April. The orbit was first computed as usual on the sup-position of its being a parabola, but the parabolic elements being found unsatisfactory, an elliptic orbit was comput-ed, and the period found to be about seven years. Its eccentricity was found to be 0·55, thus forming a connect-ing link between the asteroids and comets. Hitherto there was a well-marked distinction between planets and comets. The most eccentric of the planetary orbits is that of Polymnia (0·337) or about *one third*. The least eccentric cometary orbit hitherto well established, was that of Biela's comet, (0·75), or almost exactly *three quar-ters*. Faye's comet, with an eccentricity of *one half* (0·55) occupies an intermediate rank, and nearly removes what

had hitherto been regarded as one of the most distinctive
features of comets.

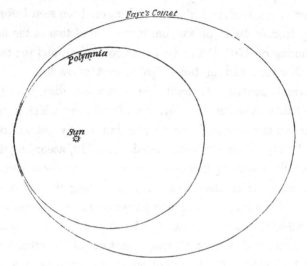

The position of its orbit in the heavens is very unstable.
At aphelion, its distance from the sun is five hundred and
sixty millions of miles, bringing it into close proximity
to the planet Jupiter. Such a conjunction happened in
1840, at which time the attraction of Jupiter for the comet
was about one tenth part of the sun's, and must have produced a considerable alteration of its orbit. In 1815, the
comet probably came still nearer to Jupiter, by which its
former orbit must have been greatly changed. The former path of this comet may, therefore, have been very
different from that which it pursued in 1843, and M. Valz
expressed the opinion that this comet was identical with
the famous comet of 1770. The latter comet, at its ap-

pearance in 1770, was found to be moving in an ellipse,
whose period was only five and a half years, and astronomers were surprised that it had never been seen before.
By tracing back its motion, it was found that at the beginning of 1767, it was very near to Jupiter, and the two
bodies remained in the neighborhood of each other for
several months.   Computation, moreover, disclosed the
fact that previous to 1767, the elliptic orbit which it described corresponded not to five, but to fifty years of revolution round the sun.   Again, in 1779, according to
Lexell's elements, it was 500 times nearer to Jupiter than
to the sun; so that then, notwithstanding the immense
size of the sun, its attractive power on the comet was not
the 200th part of that exerted by Jupiter.   It was found
that on the departure of this comet out of the attraction
of Jupiter in 1779, its circuit could not be performed in
less than twenty years.   Thus the action of Jupiter
brought the comet of 1770 to us in 1767, and removed it
from us in 1779.   Moreover, at every revolution, the
comet of 1770 ought to come into close proximity to
Jupiter, and suffer enormous perturbations, and M. Valz
conjectured that the orbit of this comet had been at last
transformed into that of the comet observed in 1843.   M.
Le Verrier has undertaken a thorough investigation of
this question, and he thinks he has demonstrated that
the comet of 1770 has nothing in common either with
Faye's comet of 1843, or with that discovered by De
Vico in 1844, or any other comet whose orbit has been
computed.

M. Le Verrier computed the perturbations of the comet arising from the attraction of the planets during the interval from 1843 to 1851, and predicted that this body would return to its perihelion on the 3d of April, 1851.

The comet was seen at the Cambridge observatory, in England, on the 25th of December, 1850, and was followed until the 4th of March. It was described as an extremely faint object, so as to be barely visible. The positions assigned to it, scarcely differed at all from those assigned in the Ephemeris computed from the elements of Le Verrier. The comet was discovered by Mr. Bond, at the Cambridge (Mass.) observatory, on the 1st of January, 1851.

Mr. Bond describes the comet as being at that time a very faint object in the twenty-three feet refractor, and as appearing slightly elongated in the direction of the sun. The same body was discovered at the Pulkova observatory, on the 24th of January.

Faye's comet is the *fourth* which has been observed to return to the sun in conformity to a prediction. Halley's comet made its first predicted return in 1759 ; Encke's comet in 1822 ; and Biela's comet in 1832. Several other comets have been predicted to return, but these predictions have not been verified by observation.

The comet of Faye may be expected to arrive at perihelion again in October, 1858.

SECTION III.

DE VICO'S COMET OF 1844.

On the 22d of August, 1844, Father De Vico, director of the observatory at Rome, discovered a telescopic comet in the constellation of the Whale. He immediately announced the discovery to Professor Schumacher, of Altona, but his letter did not arrive till the 26th of September. Meanwhile, the comet had been discovered independently by several different observers. It was seen by Professor Encke at Berlin on the 5th of September, and on the 6th it was seen at Hamburg by M. Melhop, an amateur astronomer. On the 10th of September it was discovered by Mr. H. L. Smith of Cleveland, Ohio, who observed it every day for nearly a fortnight. About the third week in September, it was just discernible with the naked eye, and with slight optical aid had a very beautiful appearance, the nucleus being bright and star-like, and having a tail about one degree in length, extending in a direction opposite to the sun. At the Pulkova observatory, the comet was followed till the 31st of December.

It was soon found by M. Faye and others, that the comet deviated remarkably from a parabolic orbit; and it was ascertained that the curve described was an ellipse with a periodic time of about five and a half years. Dr. Brünnow (formerly of Berlin, but now director of the observatory at Ann Arbor, Mich.), undertook a thorough investigation of all the observations, embracing a period of more than four months, and took account of all the planets within the orbit of Uranus. He thus obtained an orbit which satisfied all the observations with extreme precision, and indicated that the length of the comet's revolution was 1996.5 days, or 5.4659 years. This period (supposing its orbit undisturbed) would bring the comet back to perihelion about the 20th of February, 1850; but it happened, very unfortunately, that when the comet was near enough to the earth to be otherwise discerned, it was always lost in the sun's rays; the geocentric positions of the sun and comet at perihelion being nearly the same, and continuing so for some months, on account of the apparent direct movement of both bodies.

At the next visit, in the summer of 1855, it was supposed that the comet would be more favorably located in the heavens, and astronomers looked forward with great interest for its reappearance. Dr. Brünnow calculated that it would be in perihelion on the evening of August 6th, 1855. Only one observation of this body has been reported from any part of the world. On the 16th of May, M. Hermann Goldschmidt, of Paris, while searching for the comet, found a nebulous body whose right ascension was 21h. 41m. 45s.; declination 15° 38′ south. It was faint, but well seen with an object-glass of 30 lines aperture, and a magnifying power of 35. Its outline was ill defined, of irregular form, and without tail.

The above position does not differ much from that which De Vico's comet should have occupied if it had passed its perihelion about five days later than that computed by Dr. Brünnow, and M. Goldschmidt concluded that he had obtained an observation of De Vico's comet. But, according to computation, the comet was much nearer the earth on the 1st of August than it was in May, and its position in the heavens was more favorable to its visibility, so that if this comet was really seen in May, it is difficult to understand why it was not somewhere seen at a later period. Dr. Brünnow is satisfied that this supposed observation was a mistake, and that De Vico's comet was no where seen in 1855. He accounts for this failure to find the comet by the faintness of the object and the uncertainty of the ephemeris, owing to the difficulty of determining the time of perhelion passage.

Messrs. Laugier and Mauvais, of Paris, computed the orbit of the comet of 1585 from Tycho's and Rothmann's observations, and obtained elements very similar to those of De Vico's comet. They, therefore, concluded that the comet of De Vico was identical with the comet of 1585, and possibly, also, with those of 1743, 1766 and 1819.

M. Le Verrier has undertaken, by a computation of the perturbations, to decide whether this comet has been observed at any former return to the sun. He concludes

that this comet can not be identical with the famous
comet of 1770, nor with that of 1585, nor with any other
on record, unless that of 1678.   He has shown that there
is strong reason to believe that the comet of De Vico is
identical with one observed in 1678 by La Hire, and recorded in the Histoire Celeste of Lemonnier.   Dr. Brünnow has undertaken similar computations, and has arrived
at the same results as Le Verrier.

SECTION IV.

BIELA'S COMET.

On the 27th of February, 1826, M. Biela, an Austrian officer, discovered a comet; and on computing its elements, it was found that the same body had been observed in 1805 and in 1772. It was soon discovered that the comet made its revolution round the sun in a period of six years and two thirds. It was of course predicted that the comet would return in 1832. Computation also disclosed another fact, which excited no little alarm. It was predicted that on the 29th of October, 1832, the comet would cross the plane of the ecliptic at a distance of less than 20,000 miles from the earth's path. Now the comet's radius, from observations in 1805, had been determined to be greater than 20,000 miles; from which it followed, that a portion of the earth's orbit would be included within the nebulosity of the comet. It was found, however, that the earth would not arrive at this point of its orbit until a full month afterward. There was, therefore, no great danger of collision; nevertheless, no little alarm was experienced by those not much conversant with astronomy. The comet returned at the time predicted, and was observed by Sir John Herschel;

but it was extremely faint, and could only be seen in good telescopes.

In 1839, this comet must have returned again to the sun; but its position was most unfavorable for observation, and it is not known to have been observed at all.

In 1846, this comet returned to its perihelion under circumstances more favorable for observation. It was first seen at Rome on the 26th of November, 1845; at Berlin on the 28th; at Cambridge, England, on the 1st of December, and afterward at most of the observatories of Europe. It continued to be observed until the 27th of April, 1846.

This return of Biela's comet will always be remarkable in its history for an appearance quite new in the annals of modern astronomy. When first observed through the five-inch refractor at Yale College, December 29th, it was seen attended by a faint nebulous spot, estimated to be rather more than a minute of space distant from its brightest point. This surprising phenomenon was first publicly announced by Lieut. Maury of the Washington observatory. On the 13th of January, Lieut. Maury discovered, that instead of being, as usual, a single comet, it apparently consisted of two comets moving through space side by side. Each body had all the characteristics of a telescopic comet, being gradually condensed toward the center, without any well-defined disc; each being elongated on the side opposite the sun. For convenience of description, we will designate one of these bodies by the name of Biela, and the other as his companion.

There are three distinct circumstances worthy of attention—their relative *magnitude, intensity of light,* and *distance* from each other.

On the 13th of January, the companion was estimated to have one eighth the magnitude of Biela; from this time it steadily increased until the middle of February, when it was judged to be equal to Biela; after which it rapidly declined, until it disappeared during the month of March.

Again, on the 13th of January, the light of the companion was estimated to have one fourth the intensity of Biela; from which time its light increased till the middle of February, when it was judged to be somewhat brighter

APPEARANCE OF BIELA'S COMET.

March 30, 1846.   Feb. 18, 1846.   Dec. 29, 1845.

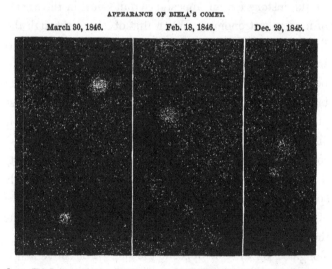

than Biela; after which it rapidly declined, and on the 1st of March could with difficulty be seen, although Biela remained quite conspicuous. At the close of March the

comet appeared single, and so continued until the 27th
of April, when it entirely disappeared.

Also on the 13th of January, the distance of the companion from Biela was estimated at about two minutes of
space; near the middle of February it was five minutes;
on the first of March the distance was ten minutes; and
at the close of March it was fifteen minutes.

The preceding facts afford abundant materials for speculation. What was the relation of these two bodies?
The appearances, when first noticed, suggested the idea
that the companion was a *satellite* to Biela. Such an idea
is now inadmissible. It is found that all the observations
are very well represented by supposing that each nucleus
described an independent ellipse about the sun. Professor Hubbard, of the Washington observatory, has computed the orbits, and finds that all the observations of
each nucleus may be represented by an elliptic orbit within
the probable limits of the errors of such observations;
from which we must conclude that the disturbing influence of one nucleus upon the other must have been
extremely small, and the observations do not appear to
be sufficiently precise to render this influence in any degree sensible. The following table shows the distance
of one nucleus from the other, in a straight line, expressed
in English miles according to the elements of Professor
Hubbard:

Distance.	Distance.
1845, Dec. 1, 117,000 miles,	1846, Feb. 17, 226,000 miles.
1846, Jan. 20, 198,000 "	25, 216,000 "
28, 213,000 "	March 5, 202,000 "
Feb. 9, 227,000 "	21, 170.000 "

Thus it appears that these two bodies, during the entire period of their visibility, remained at a distance from each other less than the mean distance of the moon from the earth; nevertheless, the perturbations arising from their mutual attraction, did not change their positions beyond a very small number of seconds at the utmost, from which we must infer that their masses were excessively small.

Biela's comet reappeared August 25, 1852, and continued visible till the 28th of September. The changes of relative brilliancy of the two comets were similar to those observed in 1846. For convenience of distinction we will call the most southern comet A, and the most northern B. The comet A was observed from August 25th to September 25th. The comet B was not noticed until September 15th and was followed until September 28th. On the 15th of September, A was fainter than B; on the 19th A was brighter than B; on the 20th both comets had the same brightness; while on the 23d and 25th A was fainter than B.

The apparent distance of one comet from the other during the whole time of their visibility was about half a degree. The absolute distances of the comets from each other according to the computation of Professor D'Arrest were as follows:

Distance.		Distance.	
1852, Aug. 27, 1,501,000 miles.		1852, Sep. 16, 1,616,000 miles.	
31, 1,531,000 "		20, 1,624,000 "	
Sep. 4, 1,559,000 "		24, 1,625,000 "	
8, 1,582,000 "		28, 1,619,000 "	
12, 1,601,000 "			

Thus it appears that during the entire period of the observations in 1852, the distance of the two bodies from each other remained nearly constant, and this distance was nearly seven times as great as the greatest distance in 1846. A comparison of the observations made in 1852 with those made in 1846, might be expected to determine the orbit of each comet with great precision. But here an unexpected difficulty presented itself. It seems impossible to decide whether what was known as the primary nucleus in 1846, is the same with the comet A or B in 1852; for all the observations are about equally well represented by either hypothesis, but there is a slight difference in favor of the hypothesis that the comet B in 1852 was the same as what we have called the primary nucleus in 1846.

What then is the history of these two bodies? Are they both new comets—is one a new comet and the other Biela—or are both of them parts of Biela?

We have seen that the distance of the two comets from each other in 1852 was more than one and a half millions of miles, while in 1846 it was only about 200,000 miles. On the 1st of December, 1845, the distance was only 117,000 miles, and by following both bodies backward we find that about the last of September, 1844, their distance from each other was only 15,000 miles. Now the radius of the primary nucleus has been computed to be 20,000 miles; from which we must conclude that the two bodies were in contact in September, 1844, and each body was within a few hours' motion of the place where Biela's comet should be at that time, from calculations based on its path when seen in 1832. No doubt then Biela has been separated into two parts, and the separation probably took place in the latter part of the year 1844. With regard to this last conclusion there is perhaps some room for question. The extraordinary changes of brightness which the two nuclei exhibited both in 1846 and 1852 clearly indicate that the brightness of these objects does not depend merely upon their distance from the earth and sun, but upon other unknown causes. These causes might have developed sufficient brightness in the companion, at its last two returns to the sun, to render it visible to us; while at its former returns, on account of its unfavorable position, the companion was too faint to be noticed.

What has caused this separation of the comet into two portions? Was it caused by collision with some foreign body? Such a collision would have materially changed the figure of the orbit, and therefore we can not suppose it to have taken place since the observation of the comet in 1772, when it was found to be pursuing nearly the same path as at present. It is probable that in case of an encounter with some other body, both bodies would have moved on in some new orbit.

Was it caused by an explosion arising from some internal force? Forces of this kind we see in operation in our own globe, ejecting liquid mountains from the bowels of the earth. The surface of our moon bears marks of similar agency—the sun appears agitated by powerful forces, perhaps the expansion of gaseous substances. If we knew that Biela's comet was a solid body, we might easily suppose it to have been divided by some force similar to volcanic agency. But most of the matter of this body is of the rarest kind; and it may be doubted whether any part of it is in a solid state.

Was this separation caused by a repulsive force emanating from the sun? The phenomena exhibited by Halley's comet at its return to the sun in 1835, require us to admit the existence of repulsive as well as attractive forces. The effect of the sun's repulsion upon the atmosphere of the comet, would be to distort it from the spherical form, which it would assume under the attraction of the nucleus alone—to crowd the particles on the side next the sun nearer to the nucleus, and to drive those on the opposite side further from it—causing an oval form, whose length, as compared with its breadth, would be the greater, the stronger the repulsive force is supposed; and the repulsive force may be conceived to become so great as to drive the remoter particles beyond the influence of the nucleus, and carry them off into space. It is necessary, in the opinion of Sir John Herschel, to suppose that the tail is attracted by the nucleus, otherwise they would at once part company. The compound mass of the comet is therefore urged toward the sun by the difference of the total attractive and repulsive forces; and so long as the repulsive power is insufficient to separate them, they will revolve together as one body, continually elongating itself as it approaches the sun, and on the position of

whose longer axis the sun exercises a directive power, as
it would on a magnet, if it were itself magnetic ; or rather
as a positively electrified body would on a non-conduct-
ing body of elongated form having one end positively,
and the other negatively excited.

As a comet approaches the sun, a portion of its matter
appears to be converted into vapor. In this vaporization,
the two electricities might be separated, the nucleus and
tail being in opposite electrical states. If now we suppose the sun to be in a permanently excited electrical state,
we have an explanation of the repulsive force which has
been ascribed to the sun. If the repulsive force of the
sun upon the particles of the tail should overcome the
attraction of the nucleus, they must be driven off irre-
coverably. Such a separation could hardly be accomplished without carrying off some portion of the gravi-
tating matter ; and thus a new comet would be formed,
as in the case of Biela.

Sir John Herschel mentions another mode in which
the division of a comet might be effected. The oscillations of a fluid covering a central body may, under certain conditions as to the coercive power of that central
mass, cease to continue of small extent, and may increase
in magnitude beyond any limit which analysis is capable
of assigning, even to the extent of destroying the con-
tinuity of the fluid, and separating it into distinct masses.
If such an extreme case could ever occur, it must be in a
comet like Biela's, consisting of a mass of vapor with

very little cohesion, and in which the attractive power of
the nucleus is exceedingly small.

Professor Santini, by employing Plantamour's elements,
had computed that this comet would arrive at its perihelion in 1852, Sept. 28.72, Berlin mean time. An error,
however, has been detected in Plantamour's computations,
which being corrected, reduces the time of perihelion
passage to Sept. 25.26. The mean time of passage of the
two nuclei was found to be Sept. 23.36 ; that is, the comet
reached its perihelion nearly two days earlier than the
time computed from observations at the preceding return.
Encke's comet exhibits a similar peculiarity, and this fact
is accounted for by supposing that a thin ethereal medium pervades the planetary spaces, sufficiently dense to
produce some impression upon a comet, but incapable of
exercising any sensible influence on the movements of
the planets.

The Imperial Academy of Sciences of St. Petersburg,
has proposed the theory of Biela's comet as a prize problem, the memoirs to be presented on the 1st of August,
1857. The Academy demands a rigorous investigation
of the elements of the orbit described by the center of
gravity of the comet, founded on a discussion of all the
observations from 1772 to 1852.

SECTION V.

MISS MITCHELL'S COMET.

THIS comet was discovered on the 1st of October, 1847, by Miss Maria Mitchell, of Nantucket. As a relaxation from the severer toil of a systematic course of observations, she had employed the intervals through the preceding year in sweeping for comets; but her labors had hitherto been only rewarded by a familiarity with comet-resembling nebulæ, which she had constantly and carefully recorded. The instrument employed on these occasions was a forty-six inch refractor, with an aperture of three inches, mounted on a tripod, and furnished with a terrestrial eye-piece of moderate power. On the evening of October 1st, a circular nebulous body appeared in the field of the telescope, a few degrees above Polaris. There was scarcely a doubt of the cometary character of this object, inasmuch as the region which it occupied had frequently been examined. Still, as the object was faint, and the weather uncommonly clear, a possibility existed that this too was a nebula not before observed. On the evening of the 2d, its change of place was manifest. No appearance of condensation of light toward its center, nor any indication of a train, could be detected.

It is evident that its apparition, even to the telescope, was sudden. Its first apparent motion was inconsiderable, and the region of its discovery had been constantly swept over by the assistant observer at Cambridge, with his excellent comet-seeker, even as late as the previous evening. This idea is strengthened by the subsequent rapid increase of the brilliancy of the comet, and the acceleration of its apparent motion. On the third, its motion and brightness had much increased, and there was noticed a slight increase of light toward its center. On the 4th, all observations were prevented by the weather. On the 5th the evening was delightful. At an early period it was evident that the comet must pass over a fixed star of the fifth magnitude, and preparations were made to note the beginning and the end of the transit; but the border of the comet proved too uncertain to rely upon. At 10h. 54m., the star appeared to be exactly in the center of the comet; and during several seconds it was impossible to determine, with a power of 100, in which direction was the greatest extent of nebulosity. It appeared in fact like the nucleus of the comet shining through it with undiminished brilliancy.

On the 6th the comet was visible to the naked eye, and continued to increase in brightness till obscured by the light of the moon. On the 9th, as seen at Cambridge, it exhibited a faint train, a degree and a half in length, and opposite to the sun.

This comet was discovered by M. De Vico, at Rome, on the 3d of October; it was discovered by Mr. Dawes,

of England, on the 7th; and on the 11th it was discovered by Madame Rümker, of Hamburg.

As there was no doubt of Miss Mitchell's having been
the first discoverer of this body, she seemed fairly entitled
to the gold medal offered by the King of Denmark for
the first discovery of a comet. In consequence, however,
of her not having complied strictly with the conditions
of giving immediate notice of the discovery by letter to
Professor Airy, it was for a time doubtful whether the
medal would not be awarded to M. De Vico. A full
statement of the circumstances of the discovery having
been made to the King of Denmark, his majesty ordered
a reference of the case to Professor Schumacher, who
reported in favor of granting the medal to Miss Mitchell.
This report was accepted by the king, and the medal has
been transmitted accordingly. This is the first instance
in which the gold medal, founded by the King of Denmark in 1831, for the first discovery of a comet, has been
awarded to an American, and the first instance in which
it has been awarded to a lady in any part of the world.

Miss Mitchell has since been provided with a comet-
seeker of a very large aperture, the manufacture of Mr.
Fitz of New York, and the public will therefore look for
additional discoveries from the same quarter.

SECTION VI.

ONE of the most splendid comets mentioned in history is that which made its appearance in the middle of the year 1264. It is recorded in terms of astonishment by nearly all the historians of the age; no one then living had seen any to be compared with it. It was at the height of its splendor in the month of August, and during the early part of September. When the head was just visible above the eastern horizon in the early morning sky, the tail stretched out past the mid-heaven toward the west, or was fully 100° in length. Both Chinese and European writers testify to its enormous magnitude. In China, the tail was not only 100° long, but appeared curved in the form of a saber. It continued visible until the beginning of October, historians generally agreeing in dating its last appearance on the 2d of October, or on the night of the death of Pope Urban IV., of which event it seems to have been considered the precursor. Rough approximations to the elements of this comet were attempted by Mr. Dunthorne in the middle of the last century, and subsequently by M. Pingré, the well-known French writer on the history of comets.

In the year 1556 another splendid comet made its ap-

7*

pearance. It was seen in some places near the end of February, and was equal in size to half the moon. Its beard was short, and was unsteady. It exhibited a movement like that of a flame, or a torch disturbed by the wind. The length of its tail was about four degrees; its color resembled that of Mars, but somewhat paler. On the 12th of March it had reached a north declination of 42°, and it moved over 15° of a great circle in a day. It was then distant from the earth only about seven millions of miles, and showed considerable train. It continued visible until the 23d of April, when it disappeared in consequence of its proximity to the sun. Dr. Halley, the second astronomer royal of England, computed the elements of this comet; but owing to the imperfect nature of the observations, his elements were not considered very exact. Mr. Dunthorne's results for the comet of 1264 were so similar to those which Halley had given for the comet of 1556, that he was immediately led to conclude these two bodies to be identical, and the period being probably about 292 years, he surmised that a re-appearance might be expected about 1848.

About twenty years later, M. Pingré collected together all the accounts he could find relative to the comet of 1264, and the result of an elaborate investigation was that the paths of the comets of 1264 and 1556 might be represented with tolerable accuracy by elements very closely similar; and hence he regarded those bodies as identical, and coincided with Mr. Dunthorne in anticipating its re-appearance about the year 1848.

Between the years 1843 and 1847, Mr. Hind of London investigated the question of identity anew, and he concluded that the comets of 1264 and 1556 were very probably the same, and that a return to perihelion might be expected about the middle of the nineteenth century.

M. Bomme, of Middleburg, in the Netherlands, has computed the effects that will be produced on the period of the comet's return by the united attraction of the planets. With Dr. Halley's elements it was found that at the time the comet was visible in 1264, it was moving in an ellipse, with a periodic time of 112,469 days, or 308 years; but perturbations, owing to planetary attraction, quickened the return of the comet by no less than 5903 days, so that it was in perihelion in April 1556, and at that time the revolution corresponding to the elliptic arc described was 112,943 days. Starting again from this epoch, M. Bomme ascertained that the next revolution should occupy 111,146 days, bringing the comet to its perihelion on the 22d of August, 1860, at which time the revolution will be 113,556 days.

Employing Mr. Hind's elements, it was found that in 1264 the ellipse described by the comet had a period of 110,644 days, or 302.922 years, and that planetary disturbances expedited its return by 4077 days. At the time it was visible in 1556, its mean motion corresponded to a period of 308.169 years. The present revolution should be shortened by perturbations 3828 days, or 10.48 years, and the comet should again reach its perihelion on the 2d of August, 1858, the revolution belonging to the major axis at that epoch being 308.784 years.

Hence it appears that there is an uncertainty of two years (between August 1858 and August 1860) in the time of the next arrival at perihelion, according as the elements of Dr. Halley or those of Mr. Hind are employed.

It is an important circumstance as bearing on the question of identity of the comets of 1264 and 1556, that about 289 years before the former epoch, or A.D. 975, a comet of great apparent magnitude was visible, which may possibly have been the same. It exhibited a tail 40° in length. Assuming the elements obtained by Mr. Hind for the comet of 1556, and that the perihelion passage took place early in August 975, it is found that the observed track may be very closely represented.

It is, therefore, inferred that the next perihelion passage of the comet of 1264 will take place within two years of August 1858; nearer than this it does not appear possible to approximate in the present state of our knowledge.

SECTION VII.

THE GREAT COMET OF 1853.

THE great comet of 1853 was discovered by M. Klin-
kerfues at Göttingen on the 10th of June, at which time
it was a somewhat faint telescopic object. About the
7th of August it began to be faintly visible to the naked
eye; on the 20th it was equal to a star of the third or
fourth magnitude; on the 25th it was equal to a star of
the second magnitude; on the 30th it was as bright as
one of the brightest stars of the first magnitude. From
the 30th of August to the 4th of September, although
the comet was distant but a few degrees from the sun's
place, it was seen and well observed each day in full day-
light by M. Schmidt of Olmütz. On the 31st of August
he made a series of observations of the comet's place
with his telescope about mid-day, although the comet
was only twelve degrees from the sun; and on the 2d,
3d and 4th of September, he observed it at mid-day,
although only seven or eight degrees from the sun. Also
on the 3d of September, about noon, Mr. Hartnup, of the
Liverpool observatory, saw the comet distinctly with his
telescope.

For a month after the discovery of this comet, it exhibited an oval form, its greatest diameter being about three minutes. After this date it became more elongated, and by the first of August had attained a length of ten or fifteen minutes. On the 20th of August it exhibited a tail about one degree in length; and during the next ten days the tail rapidly increased, attaining at last a length of about fifteen degrees.

The comet passed its perihelion on the 1st of September; and it is not known to have been seen any where in the northern hemisphere after the 4th of September. On the 16th of September the comet was noticed at Santiago de Chili, and was observed there till the 7th of October. It was visible to the naked eye, its nucleus being very well defined; but its tail was so faint, that it was scarcely possible to determine its length. At the Cape of Good Hope the comet was seen and carefully observed from the 11th of September until the 11th of January, 1854. It was also observed at New Zealand from the 14th of September to the 10th of October. On the 15th of September the length of its tail was about five degrees, from which time its brightness daily diminished.

All the observations of this comet are tolerably well represented by a parabolic orbit.

CHAPTER III.

ADDITIONS TO OUR KNOWLEDGE OF THE FIXED STARS AND NEBULÆ.

SECTION I.

DETERMINATION OF THE PARALLAX OF FIXED STARS.

UNTIL recently, astronomers had been unable to measure the distance of a single fixed star. The parallax arising from the motion of the earth in its orbit, even for the nearest fixed star which had been examined, remained concealed among the small errors to which all astronomical observations are liable. Nevertheless, it was generally agreed among astronomers that no star visible in northern latitudes, to which attention had been directed, manifested an amount of parallax exceeding a single second of arc. An annual parallax of one second implies a distance of about *twenty millions of millions of miles*, a distance which light, traveling at the rate of 192,000 miles per second, requires $3\frac{1}{4}$ years to traverse. This being the inferior limit which the nearest stars exceed, it is not unreasonable to suppose that among the innumerable stars which the

telescope discloses, there may be those whose light requires hundreds, and perhaps thousands of years to travel
down to us.

The difficulty of measuring, by direct meridional observations, a quantity so minute as the parallax of the
stars, has led astronomers to try a system of differential
observations, susceptible of far greater accuracy. Suppose
there are two stars at unequal distances from us, so situated as to appear nearly on the same line of vision. Their
apparent places must be alike affected by aberration, pre-
cession, nutation, refraction and instrumental errors; so
that although it is difficult to determine the true right
ascension and declination of either star within one second
of arc, we may measure the *difference* of position of one
star from the other with extreme precision, without the
necessity of taking account of the preceding corrections.
Now the difference of position of the two stars, if measured for every season of the year, gives us their *difference
of parallax;* and if one star is several times more distant
than the other, this difference of parallax will be sensibly
the entire parallax of the nearer star. This method was
first proposed by Galileo more than two centuries ago,
but it does not appear that he ever attempted to reduce it
to practice. Sir William Herschel attempted to reduce
this method to practice, and for this purpose he selected a
great number of double stars which in consequence of the
inequality of the component members, appeared to be well
adapted to this object. His labors led to the discovery
of a physical connection between the bodies composing

double stars, but he did not succeed in establishing any parallax.

In the year 1837, the late Professor Bessel applied this method to determine the parallax of the double star 61 Cygni. This is a small star, hardly exceeding the sixth magnitude, which had been pointed out for special observation by the circumstance of its having a very large proper motion, viz., more than five seconds per year, being a more rapid motion than had been detected (until recently) in the case of any other star, on which account it had been suspected to be comparatively near our system. Bessel repeatedly measured the distance of this star from two other stars in its neighborhood, both of them very minute, and therefore presumed to be very distant: and he continued his observations every month, when practicable, for three years. The distances were measured by means of a magnificent heliometer, which Fraunhofer of Munich, had recently executed for the observatory of Königsberg. It appeared that in January of each year, the distance of 61 Cygni from one of the stars of comparison was one third of a second *less* than the mean distance; while in June it was one third of a second *greater* than the mean. This effect is precisely such as should be produced by the motion of the earth about the sun, causing an apparent displacement of the nearer stars as compared with those which are more remote, and as no other explanation of the phenomenon seems admissible, these observations are considered as settling the long vexed question of parallax. In 1841, Professor Bessel received the

gold medal of the Royal Astronomical Society of London, for this important discovery. The parallax of 61
Cygni, according to these observations, is 0″.348, making
the distance of this star from the sun 592,000 times the
radius of the earth's orbit—a distance which light would
require more than nine years to traverse.

This result has received confirmation from the researches of M. Peters, who from a series of zenith distances of the star, determined at the observatory of Pulkova during the years 1842 and 1843, has found its parallax to be equal to 0″.347. These observations were made
with the grand vertical circle of Ertel, which is 43 inches
in diameter, graduated to 2 minutes, and reading by four
microscopes to one tenth of a second.

A series of observations of 61 Cygni have also been
made by Mr. Johnson, with the new heliometer of the
Oxford observatory. These observations were commenced in 1852 and were continued through the years
1853 and 1854, and they fully confirm the fact of an
annual parallax, very nearly of the same amount as that
found by Bessel. The stars selected for comparison were
different from those which Bessel used, both stars lying
nearly in the direction of the components of 61 Cygni.
These observations give the parallax 0″.392 or ⁻0″.402,
according as a temperature correction is applied or not.

M. Otto Struve, of the Pulkova observatory, has made
some observations of 61 Cygni with a wire micrometer
attached to his telescope, and a preliminary calculation
has furnished for the parallax 0″.52, a result sensibly

greater than has been obtained by either of the preceding observers.

In 1839, Professor Henderson, of Edinburgh, announced that in discussing his observations of Alpha Centauri made at the Cape of Good Hope, with a mural circle, during the years 1832–3, he had found evidence of a sensible parallax. The double star, a Centauri is one of the brightest stars of the southern hemisphere. The two stars a^1 and a^2 are situated within $19''$ of space of each other. On comparing the observations of Lacaille with those of the present time, it is found that although the two stars have not sensibly changed their relative positions, each has an annual proper motion of 3.6 seconds of space. It thus appears that they form a binary system, having one of the greatest proper motions that has been observed; and from this circumstance, as well as the brightness of the stars, it is reasonable to suppose that their parallax may be sensible. Professor Henderson's observations were not made for the purpose of ascertaining the parallax, but of accurately determining the mean places of the stars. On reducing the declinations, he found a sensible parallax, but he delayed communicating the result until it should be seen whether it was confirmed by the observations of right ascension made by Lieutenant Meadows with the transit instrument. These observations also indicated a sensible parallax. A combination of all Professor Henderson's observations gave a parallax of $1''.16$, with a probable error of $0''.11$.

In 1839 and 1840 a series of observations was made by

Mr. Maclear, at the Cape of Good Hope, for the purpose of ascertaining the parallax of this star. The observations consisted of double altitudes measured with the mural circles, and they gave a parallax of $0''.91$; and from subsequent observations extending down to 1848, he found it to be $0''.98$.

Mr. Maclear also made a series of observations of β Centauri, in the years 1842–44, from which he has deduced a parallax for this star amounting to $0''.47$, with a probable error of $0''.04$.

In the year 1835, M. Struve commenced a series of observations, with a view of detecting the parallax of the bright star a Lyræ. His method of observation consisted in measuring with a micrometer the distance between this star and another very small star situated about $43''$ from it, repeating the operation at different seasons throughout the year. The result of this inquiry assigned a parallax of $0''.26$ to a Lyræ.

The more recent researches of M. Peters at the Pulkova observatory tend to prove that this star has a visible parallax; although the result is less than that assigned by M. Struve. These observations were made with the great vertical circle of Ertel, and his result was a parallax of $0''.10$, with a probable error of $0''.05$.

M. Otto Struve has recently given the result of fifteen months' observations on this star, and finds a parallax of $0''.14$. Combining this result with the former values of M. Struve and Peters, he obtains a mean value of $0''.15$, with a probable error of $0''.01$.

The star 1830 of Groombridge's catalogue (right ascension 11h. 44m. 18s., declination 38° 48′ north), has the largest proper motion at present known, exceeding 7″ in arc; and in consequence various attempts have been made at different places to determine its parallax, the probability being great that its parallax would be appreciable. The discovery of the proper motion was published by Argelander in the year 1842, and Bessel was induced to give immediate directions to Schlüter, his assistant, to commence a series of measures with the Königsberg heliometer. These measures were begun in October, 1842, and were terminated abruptly near the end of August, 1843, by the illness and subsequent death of Schlüter. M. Wichmann was then charged with the continuation of the observations, but more pressing duties prevented him from completing the observations till the year 1851.

In the year 1846 Mr. Faye, by observing with the equatorial of the Paris observatory the difference of right ascension between this star and that of another small star situated nearly on the same parallel, determined the parallax to be 1″.08, and by this striking result drew the attention of others to the star. Peters, by discussion of his observations made at Pulkova with Ertel's circle, subsequently determined its parallax to be 0″.226, with a probable error of 0″.141. This result is derived from 48 zenith distances observed in 1842 and 1843. Mr. Wichmann's discussion of Schlüter's observations gave a parallax of 0″.18. M. Otto Struve, by micrometer measures of

differences of north polar distance made with the great
refractor in 1848 and 1849, found the parallax to be 0″.03.

In the year 1852 Mr. Wichmann published the results
of his new series of observations with the heliometer
made in 1851. In these observations he measured the
distance between 1830 Groombridge and a small star a
preceding by 34′ in right ascension, a second star a' following it by 34′ in right ascension, and a third star a'' following it by 30′ in right ascension. The result of his
discussion of the observations was that the parallaxes of
the stars a' and a'' were of very small amount, but that the
parallax of 1830 Groombridge, amounted to 0″.71, and
that of the star a to 1″.17.

M. Peters has subjected the observations of M. Wichmann to a critical discussion, and has shown that Wichmann's adopted temperature correction is probably too
small.   He considers the most trustworthy result de-
ducible from the observations to be a parallax of 0″.148
for 1830 Groombridge.

During the years 1852, '53 and '54, a long series of observations of this star were made by Mr. Johnson with
the Oxford heliometer, and these also present an anom-
alous result, inasmuch as they assign a greater parallax
to one of the stars of comparison than to 1830 Groombridge; and what adds to the perplexity is, that the star
which appears to have the greater parallax in this case,
is one of those which in M. Wichmann's researches
appeared to be most distant. Mr. Johnson's results are
a parallax of 0″.26 for 1830 Groombridge, and of 0″.44

for the star of comparison a'', while the parallax of star a was considered to be insensible. These results appear to indicate some imperfection in double-image measures of objects separated by large arcs, which still remains to be explained.

Mr. Henderson has attempted to determine the parallax of Sirius from the observations made with the mural circle at the Cape of Good Hope in the years 1832 and 1833, as also in 1836 and 1837, being in all 231 observations. From these he has deduced a parallax of $0''.23$. The error of this determination he estimates not to exceed a quarter of a second, from which he concludes that the parallax of Sirius is not greater than half a second of space, and that it is probably much less.

Numerous attempts have been made to determine the parallax of the pole-star. From a comparison of 603 right ascensions of the pole-star observed by M. Struve and Preuss at Dorpat from 1822 to 1838, M. Peters has deduced a parallax of $0''.17$, with a probable error of $0''.03$. From a comparison of the declinations of the pole-star observed at Dorpat from 1822 to 1838, M. Peters has deduced a parallax of $0''.15$; and by comparing the right ascensions observed at Dorpat from 1818 to 1821, he has deduced the parallax of $0''.07$. M. De Lindenau, from a comparison of 890 right ascensions of the pole-star, observed by different astronomers, deduced a parallax of $0''.14$, with a probable error of $0''.06$. From 289 observations of the pole-star made at Pulkova in 1842 and 1843 with the grand vertical circle of Ertel, M. Peters has deduced a parallax of $0''.07$. Taking the mean of all these determinations, M. Peters has obtained as a final result the parallax of the pole-star $0''.106$, or about one tenth of a second, a distance which light would require 30 years to traverse. According to this result it must have been 30 years after the pole-star was placed in the firmament before its light shone upon the earth; and if it were now annihilated, it would still shine for 30 years longer to guide the mariner across the ocean.

M. O. Struve has recently announced that he has found the parallax of a Cassiopeæ to be $0''.34$ with a probable error of $0''.05$; and the parallax of a Aurigae to be $0''.30$ with a probable error of $0''.04$.

M. Peters has also attempted to determine the average value of the parallax of stars of the different magnitudes. For this purpose, he availed himself of the long series of observations made at Dorpat, and concludes that the average parallax of stars of the second magnitude is $0''.116$. Combining this result with the relative distances of stars of the different orders as determined by M. Struve, we are enabled to estimate the parallaxes of the stars of the various orders, and hence their absolute distances. The following is M. Peters' result for stars of the first six magnitudes:

App. magnitude.	Parallax.	Distance in radii of the earth's orbit.	Time required for light to traverse this distance.
1	$0''.209$	986,000	15 years.
2	$0''.116$	1,778,000	28 "
3	$0''.076$	2,725,000	43 "
4	$0''.054$	3,850,000	61 "
5	$0''.037$	5,378,000	85 "
6	$0''.027$	7 616,000	120 "

Thus light, which travels at the rate of 190,000 miles every second, requires 15 years to come to us from stars of the first magnitude, and 120 years to come from one of those small stars which are just visible to the naked eye. These results can, of course, only be regarded as provisional, until sufficient observations are collected for a more thorough investigation of the subject; but it is not too much to expect that ere long a similar table for at least the brighter stars may be constructed, founded on unquestionable data.

8

SECTION II.

OBSERVATIONS OF NEW AND VARIABLE STARS.

It has long been known that among the fixed stars are several which experience a periodical increase and diminution of brightness. The star *Omicron*, in the constellation Cetus, sometimes appears as a star of the second magnitude, but continues of this brightness only about a fortnight, when it decreases for about three months, till it becomes completely invisible to the naked eye, in which state it remains about five months, and then increases again to the second magnitude, the interval between its periods of greatest brightness being about eleven months. The star Algol varies from the second to the fourth magnitude, going through its changes in less than three days. More than forty such cases have been noticed, although in many of them the change of brightness is not very remarkable.

In the case of a few stars, remarkable changes of brightness have been observed, which have not been reduced to any law of periodicity. The star η (eta) Argûs is of this kind. This is a star of the southern hemisphere in right ascension 10h. 39m.; south declination 58° 54'. In Halley's catalogue, constructed in 1677, it is marked as of the fourth magnitude; yet in Lacaille's in 1751, and in subsequent catalogues, it is recorded as of the second magnitude. In the interval from 1811 to 1815, it was again of the fourth; and again from 1822 to 1826 of the second magnitude. In 1827, it increased to the first magnitude; it thence receded to the second, and so continued until the end of 1837. In the beginning of 1838, it suddenly increased in luster so as to surpass all the stars of the first magnitude except Sirius, Canopus, and Alpha Centauri, which last star it nearly equaled. Thence it again diminished, and in 1842, it was pronounced by Maclear as inferior to Alpha Crucis, but the next year it again revived, and became nearly equal to Sirius.

These facts afford abundant materials for speculation. The changes of brightness of η Argûs are spread over centuries, and apparently without any regular period. What can be the cause of these changes? To this question we are unable to assign any satisfactory answer, and must wait patiently until a greater accumulation of facts shall afford us a more certain basis for a theory.

Several instances are on record of *temporary stars*, which have suddenly become visible, and after remaining a while, apparently immovable, have died away and left no trace behind. Such a star is said to have appeared about the year 125 B. C. Such stars are also recorded in the years A. D. 389, 945, 1264, 1572, 1604, and 1670. A similar phenomenon has recently taken place. On the

27th of April, 1848, Mr. Hind, of London, observed a star of the sixth magnitude in the constellation Ophiuchus, where he was certain that, up to the 5th of that month, no star as bright as the ninth magnitude previously existed. Neither has any record been discovered of a star being there observed at any previous time. Its place was in right ascension, 16h. 51m. 1s., south declination, 12° 39′ 14″. On the 2d of May he estimated it to be of the fifth magnitude, or a little brighter, and therefore distinctly visible to the naked eye. Its light was reddish in the telescope; and Dr. Petersen observed that the reddish color at times increased suddenly in intensity, and again as suddenly disappeared. Other observers noticed these peculiar red flashes. On the 19th of May, Mr. Hind pronounced it fainter than when he first noticed it; and on the 24th, it was ranked as a star of the sixth magnitude. The Messrs. Bond, at Cambridge, made a series of comparisons between this star and one of the fifteenth magnitude in its vicinity, and found that during three months of observation its position remained unchanged. They remarked, "This star resembles Antares, but its red is deeper. It is one of the most strikingly colored stars we remember to have seen. With a power of 1500 it showed no sign of a planetary disc." On the 15th of August they state, "This star appears to have decreased in brilliancy, and is now of the seventh magnitude; its ruby red color still remains. It is at once recognized from its neighbors by its color alone." On the 23d of March, 1849, Professor Kendall, of Philadelphia, pronounced this star to be of the *eighth* magnitude. On the evening of June 4th, 1850, the writer made a careful survey of all the stars in this vicinity, and found only one which could be estimated as high as the tenth magnitude, and this had no very decided red color. Hind's star may therefore now be pronounced extinct.

Hind's star was not many degrees distant from the place where a new star was seen in 1604. This star was first announced on the 10th of October, and it was seen by Kepler on the 17th. The star was perfectly round, without nebulosity or tail; its light was brighter and more unsteady than that of the other stars. After it had risen above the vapors of the horizon, its light was white. It not only surpassed stars of the first magnitude, but also Mars and Jupiter. Some even compared it to Venus, but Kepler was not of this opinion. Observations proved that it had no motion or sensible parallax. On the 9th of November it was seen in a twilight which rendered Jupiter invisible. On the 16th, Kepler saw it for the last time, before its conjunction with the sun. On the 24th of December, it reappeared in the east with diminished brightness. It was still brighter than Antares, but inferior to Arcturus. On the 20th of March it appeared smaller than Saturn ; but it was much larger than the stars of the third magnitude in Ophiuchus. On the 13th of September it was smaller than the third magnitude, and on the 8th of October it could be seen with difficulty. A few days later it disappeared in the sun's rays. In January and February some observers thought they saw this star again, but without

being confident of it. In the month of March it could
not possibly be seen, so that it must have disappeared between October, 1605, and February, 1606. Mr. Hind's
star is about 12° north of the place of that discovered
in 1604, and about 17 minutes of time less in right ascension.

Mr. Hind has recently announced another scarlet star,
between Orion and Eridanus. Its place is in right ascension 4h. 52m. 45s., declination 12° 2′ south. He says,
"I found this star in October, 1845, and have kept a close
watch upon it since. It is of about the seventh magnitude, and the most curious colored object I have seen."

In November, 1850, Mr. Hind discovered a new star of
the seventh magnitude, of a fiery color, with a dull planetary aspect. Its place was in right ascension 1h. 22m.
54s., declination 2° 6′ north. This star is not in the
Histoire Celeste, or Bessel's Zones, nor does it occur on
the star maps of the Berlin Academy.

Mr. Hind, in the course of his observations in search
of planets, has discovered fifteen new variable stars. Probably most of these will be found to belong to the class
of periodical stars.

SECTION III.

BEFORE the invention of the telescope, it was impossible to acquire a very precise knowledge of the distance of the stars and their distribution in space; and even after the invention of the telescope, no one attempted to use it in any adequate manner to determine the constitution of the heavens, until the time of Sir William Herschel. This astronomer, having the command of instruments far superior to any who had preceded him, undertook a series of exact observations, upon which to found a knowledge of the starry heavens. In 1784 he first advanced an hypothesis respecting the Milky Way, which was substantially as follows: The stars of our firmament, instead of being scattered in all directions indifferently through space, constitute a cluster with definite limits, in the form of a stratum, of which the thickness is small in comparison with its length and breadth; and in which the earth occupies a space somewhere about the middle of its thickness, and near the point where it subdivides into two principal laminæ, inclined at a small angle to each other. For, to an eye so situated, the apparent density of the stars, supposing them pretty equally scattered through the

space they occupy, would be least in the direction of a
visual ray perpendicular to the lamina, and greatest in
that of its breadth; increasing rapidly in passing from
one to the other direction, just as we see a slight haze in
the atmosphere thickening into a decided fog-bank near
the horizon, by the rapid increase of the mere length of
the visual ray.

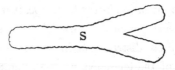

Herschel was conducted to this view of the Milky Way
by the following considerations. Supposing the stars to
be situated, in general, at equal distances from each other,
the number of stars observed in the field of a telescope
ought to be about the same in all possible directions, provided the stars extend in all directions to the same distance.   But if we have a stratum of stars at equal distances from each other, of a form whose thickness is small
in comparison with its diameter, then the number of stars
visible in the different directions, will lead us to a knowledge both of the exterior *form* of the starry stratum, and
of the place occupied by the observer.   For example, if
within a certain circle of the heavens we count ten stars,
and in a circle of the same diameter, taken in a different
direction, we count eighty stars with the same telescope,
the lengths of the two visual rays will be in the ratio of
1 to 2, or the cube roots of 1 and 8.   This is substantially
Herschel's method of *star-gages*, in which he employed a

telescope of 18 inches aperture, with a field of view of about a quarter of a degree. Herschel made 3400 gages of this kind. These gages indicate the number of stars visible in the field of his telescope, and from this number he deduced the corresponding lengths of the visual rays. Assuming the distance of the nearest of the fixed stars in accordance with the estimate of Struve, we find that, according to Herschel, the stars upon the borders of our stratum, in the constellation of the Eagle, are at such a distance as light requires 7000 years to traverse—and from the remoter stars, light would require 13,000 years to come to us.

The figure on the next page is designed to represent the form of the nebula in which our own solar system is placed, according to the views advanced by Sir William Herschel.

The great nebulæ of the heavens, such as those of Orion and Andromeda, Herschel conjectured to be Milky Ways like our own, only of much superior dimensions. The nebula of Andromeda, which he concluded to be the nearest, he placed at a distance 2000 times greater than that of stars of the first magnitude.

This hypothesis respecting the phenomena of the Milky Way, would be tenable, provided it were true, 1st, that the stars are uniformly distributed through space; and 2d, that Herschel was able, with his telescope of 20 feet, to penetrate to the limits of our stratum.

With regard to the first of these hypotheses, we find that Herschel himself subsequently abandoned it as unten-

8*

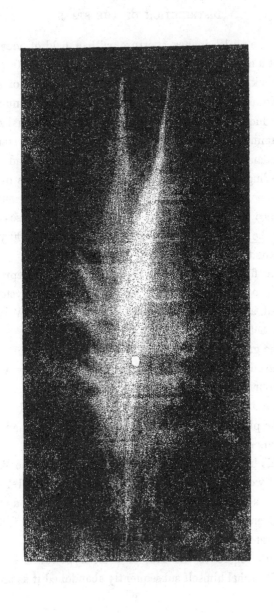

able. In 1796, he says, "The hypothesis of a uniform distribution of the stars is too far from the exact truth, to serve as a basis in this research." Again, in 1811, he adds, "The uniform distribution of the stars may be admitted in certain calculations; but when we examine the Milky Way, *this equal distribution must be abandoned.*" And in 1817 he says, "Although an increased number of stars in the field of the telescope is generally an indication of their greater distance, *my gages refer more directly to the degree of condensation of the stars.*"

As to the second of the above conditions, viz., that in his gages he was able to penetrate to the extreme limits of the Milky Way, Herschel's views underwent an entire change in the progress of his researches. In 1817, speaking of some of his gages, he says, "It is plain that the extreme penetrating power of the 20 feet telescope was insufficient to sound the depth of the Milky Way." Again, in 1818, he says, "In these ten observations the gages were arrested in their progress by the extreme faintness of the stars. There is, however, no doubt respecting the further extent of the starry region. For if in one of the observations, a feeble nebulosity had been suspected, the application of a higher magnifying power showed that the doubtful appearance was caused by the blending of numerous stars, too small to be seen by the aid of a lower magnifying power. We hence infer that if our gages cease to resolve the Milky Way into stars, it is not because its nature is doubtful, *but because it is fathomless.*"

Thus we see that the hypothesis which Herschel an-

nounced in 1785, with regard to the constitution of the
Milky Way, and which is still connected with Herschel's name in almost all the popular treatises on astronomy, was afterward *substantially abandoned by its
author.*

Quite recently, M. Struve, of Pulkova, has undertaken
a discussion of the same subject, employing, as the basis
of his researches, the most extensive catalogues of stars.
He has determined, partly by enumeration, and partly by
estimation, the number of stars of each class, as far as the
ninth magnitude, and their distribution throughout the
heavens. He finds that these stars are not uniformily
distributed; but that near the equator they are most
abundant, in the neighborhood of two points almost
diametrically opposed, viz., in right ascension 6h. 40m.,
and 18h. 40m. The stars are least abundant near the
diameter passing through 1h. 30m., and 13h. 30m. This
diameter makes an angle of 78° with the preceding. The
diameter of greatest condensation coincides almost exactly
with the position of the Milky Way; thus proving that
the phenomena of the Milky Way are intimately connected with the distribution of the stars from the first to
the ninth magnitudes, or rather that the two phenomena
are identical. Herschel proved, in 1817, that the Milky
Way was fathomless, even with his telescope of 40 feet.
The same uncertainty respecting the limits of the visible
stars exists in every part of the heavens, even toward the
poles of the Milky Way. Hence, *if we regard all the fixed
stars which surround the sun as forming one grand system,*

that of the Milky Way, we are entirely ignorant of its extent, and have not the least idea of the external form of this immense system.

The two opposite points of the celestial sphere around which the stars are observed to be most sparse, have been called the *Galactic Poles;* and the great circle at right angles to the diameter joining these points, has been denominated the *Galactic Circle.* For convenience, we will express the distance of different points of the firmament from the galactic circle, in either hemisphere, by the terms north or south *Galactic Latitude.*

The observations made in the northern hemisphere by Sir W. Herschel, and subsequently continued in the southern hemisphere by Sir J. Herschel, have supplied data for determining the law of the distribution of the stars according to their galactic latitude. The telescope used in these observations had 18 inches aperture, 20 feet focal length, and a magnifying power of 180. It was directed indiscriminately to every part of the celestial sphere, visible in the latitude of the places of observation. The observations of Sir J. Herschel were made during his residence at the Cape of Good Hope, in the years 1834–8. About 2300 gages were obtained, distributed with tolerable impartiality over the southern heavens.

An analysis of the observations of Sir W Herschel was made by Professor Struve, with a view of determining the mean density of the stars in successive zones of galactic latitude; and a similar analysis has been made of the observations of Sir J. Herschel. If we imagine the celestial sphere to be divided into a series of zones, each measuring 15° in breadth, and bounded by parallels to the galactic circle, the average number of stars included within a circle 15′ in diameter will be that which is given in the second column of the following table.

Galactic Latitude.	Average number of stars in a circle 15′ in diameter.
90°—75° North.	4.32
75°—60° "	5.42
60°—45° "	8.21
45°—30° "	13.61
30°—15° "	24.09
15°— 0° "	53.43
Galactic Circle	122.00
0°—15° South.	59.06
15°—30° "	26.29
30°—45° "	13.49
45°—60° "	9.08
60°—75° "	6.62
75°—90° "	6.05

It appears, therefore, that the variation of density of the visible stars, in proceeding from the galactic poles, is subject to almost exactly the same law of decrease in either hemisphere; the density being, however, somewhat greater in the southern than in the northern hemisphere.

M. Struve has attempted to form an estimate of the relative distances of stars of the different magnitudes as deduced from their number, supposing the stars to be distributed at uniform distances from each other along the middle of the Milky Way, and obtained the following results:

Magnitude of stars.	Mean distance.	Magnitude of stars.	Mean distance.
1	1·0000	6	7·7258
2	1·8031	7	11·3262
3	2·7639	8	19·6405
4	3·9057	9	31·2904
5	5·4545	12	227·7820

According to this table, the distance of stars of the sixth magnitude, which are just visible to the naked eye, is about eight times as great as that of stars of the first magnitude. The distance of stars of the twelfth magnitude, by which is meant those stars which were barely visible in Herschel's twenty feet telescope, is 228 times as great as that of stars of the first magnitude.

Assuming the preceding relative distances, and also that the average parallax of a star of the second magnitude is $0''.116$, we obtain the absolute distance of stars of each magnitude as given on page 168.

Professor Encke, of Berlin, has criticized these speculations of Struve with some severity. He thinks that they involve several hypotheses which are altogether unwarrantable. They assume,

I. That the apparent brightness of the stars is the simple effect of distance, so that we can assign the radius of the sphere within which the stars of each class are comprised. Encke objects to this assumption, 1. That it is contrary to the analogy of our solar system, in which the magnitudes of the planets are very unequal.

2. It is contradicted by the parallax of the stars so far as the same has been determined. The parallax of several stars of the fifth and sixth magnitudes is greater than that of most stars of the first magnitude. The star 61 Cygni, of the fifth magnitude, has a parallax certainly greater than what Struve has given for the average parallax of stars of the first magnitude. 3. It is contradicted by the phenomena of the binary stars. There is a large number of double stars, which are proved to be physically connected, and therefore both are situated at nearly the same distance from the earth, while in most of them there is a perceptible disparity of brightness, and in some cases this disparity amounts to four magnitudes.

II. The distribution of the stars over the entire heavens is admitted not to be uniform, nevertheless such a uniform distribution is assumed for the plane of the Milky Way, and the irregularities which we observe in it are regarded as in part unimportant, and in part ascribed to the eccentric position of the sun, and its distance from the plane of the Milky Way. Hence it is inferred that from the number of the stars of a given brightness, we may determine the ratio of the radius of their sphere to that of stars of any other brightness.

These and several other hypotheses which are involved in Struve's reasoning, Professor Encke regards as altogether inadmissible, and he concludes that the mean parallax which Struve ascribes to stars of the first mag nitude (viz. $0''.209$) is entirely unworthy of confidence.

Sir J. Herschel has computed from his gages, that the number of stars visible enough to be distinctly counted in the twenty feet reflector, in both hemispheres, is about *five and a half millions*. That the actual number is much greater than this, he infers from the fact that there are large tracts of the Milky Way so crowded as to defy counting the gages, not by reason of the smallness of the stars, but on account of their number.

Sir J. Herschel, in counting the gages, not only set down the total number of stars, but the number for each magnitude down to the eleventh, and even for the estimated half-magnitudes. Upon classifying the stars according to their magnitudes, it appears that the increase of density in approaching the Milky Way is quite imperceptible among stars of a higher magnitude than the eighth, and except on the very verge of the Milky Way itself, stars of the eighth magnitude can hardly be said to participate in the general law of increase. For the ninth and tenth magnitudes the increase, though unequivocally indicated over a zone extending at least 30° on either side of the Milky Way, is by no means striking. It is with the eleventh magnitude that it first becomes conspicuous, though still of small amount when compared with that which prevails among the mass of stars inferior to the eleventh, which constitute *sixteen seventeenths* of the totality of stars within thirty degrees on either side of the Milky Way.

From these observations Sir J. Herschel draws the two following conclusions; viz.—" 1st. That the larger stars

are really nearer to us (taken *en masse*, and without deny-
ing individual exceptions) than the smaller ones.   Were
this not the case, were there really among the infinite
multitude of stars constituting the remoter portion of the
galaxy, numerous individuals of extravagant size and
brightness, as compared with the generality of those
around them, so as to overcome the effect of distance, and
appear to us as larger stars, the probability of their occurrence in any given region would increase with the
total apparent density of stars in that region, and would
result in a preponderance of considerable stars in the
Milky Way, beyond what the heavens really present over
its whole circumference.   2d. That the depth at which
our system is plunged in the sidereal stratum constituting
the galaxy, reckoning from the southern surface or limit
of that stratum, is about equal to that distance which, on
a general average, corresponds to the light of a star of
the ninth or tenth magnitude, and certainly does not
exceed that corresponding to the eleventh."

This last conclusion seems clearly to assume, not only
that our Milky Way consists of a stratum of stars which has
determinate limits, but that these limits (at least in certain
directions) have been indicated by observation.   It does
not, however, appear that such a conclusion is authorized
by the gages.   The number of stars of the smallest magnitude visible, increases rapidly even up to the poles of
the Milky Way, although this increase is most rapid in
the middle zone of the galaxy.   The gages therefore indicate a condensation of stars in the neighborhood of the

Milky Way, but not that our telescopes have penetrated to the *boundaries* of our stratum. Every increase in the power of our telescopes has hitherto disclosed new stars in every part of the heavens; it is therefore unphilosophical to infer that such would not continue to be the case if we could command a further increase of telescopic power. It is, however, remarkable that those portions of the heavens which are most remote from the Milky Way are richest in nebulæ and clusters of stars. In the neighborhood of the north pole of the Milky Way, within a region occupying about one eighth of the whole surface of the sphere, one third of the entire nebulous contents of the heavens are congregated. A large portion of these nebulæ have been resolved into clusters of stars, and these stars, upon the principle that faintness is merely the effect of distance, must be inferred to be as near to us as the faintest stars of the Milky Way.

On the whole, we must conclude that the stars, *in every part of the heavens*, extend to a distance beyond the reach of the most powerful telescope hitherto constructed; that therefore the *shape* of that portion of space which the stars occupy is entirely unknown to us; that within this space the stars are not uniformily distributed, and are most crowded in the neighborhood of a plane which we call the Milky Way; that out of this plane the stars exhibit a great many centers of attraction, about which an immense number of them are clustered; but that the entire space, so far as we can perceive, is studded, though more sparsely, with stars. The material universe therefore appears to us *boundless ;* and astronomers, certainly of the present age, need not be apprehensive that they will ever witness the time when there will be no more worlds to conquer.

SECTION IV.

MOTION OF THE SUN AND FIXED STARS.

To common observation, the fixed stars retain sensibly the same relative position from age to age; but the exact observations of modern astronomy have detected a relative motion in a large number of them. A small star in the leg of the Great Bear (called 1830 Groombridge) has an annual motion of seven seconds of arc as compared with neighboring stars; and there are more than thirty stars known, whose annual proper motion exceeds one second. It might have been expected, *a priori*, that motion of some kind must exist among such a multitude of objects all subject to mutual attraction; and it appears highly probable, not to say certain, that the sun must participate in this movement. The effect of a motion of the sun, with reference to the stars, would be an apparent divergence or separation of those stars toward which we were moving, and an apparent convergence or closing up of the stars in the region which we were leaving. We might therefore expect to detect such a motion, if it really exists, by comparing the proper motions of all the stars in the firmament. In accordance with this idea, Sir William Herschel, in 1783, by a comparison of the

proper motions of such stars as were then best ascertained, arrived at the conclusion that the sun had a relative
motion among the fixed stars, in the direction of a point
in the constellation Hercules, whose right ascension is
260° 34', and declination 26° 17' north.

More recently, M. Argelander, by comparing the proper
motions of 390 stars, has located this point in right ascension 257° 35', and declination 36° 3' north.   M. Luhn-
dahl, by a comparison of the proper motions of 147
stars, has obtained for this point, right ascension, 252°
53', declination 14° 26' north; and M. Struve, by comparing the proper motions of 392 stars, has located this
point in right ascension 261° 22', declination 27° 36'
north.   The most probable mean of the results of these
three astronomers is right ascension 259° 9', declination
34° 37' north, which it will be seen does not differ greatly
from the point originally assigned by Sir W. Herschel.

Quite recently Mr. Galloway has made a similar comparison of stars visible in the southern hemisphere; and
from the proper motion of 81 southern stars not employed in the preceding investigations, he has located
the point toward which the sun is moving, in right ascension 260° 1', declination 34° 23' north, a result almost
identical with that obtained in the northern hemisphere.

It seems then nearly certain that the apparent motion
of these stars is due, at least in part, to a relative motion
of our sun, and the same observations afford us the means
of estimating its velocity.   According to Struve's calculations, this velocity is such as would carry it annually

over an angle of one third of a second, if seen at right angles from the average distance of a star of the first magnitude. If we assume the parallax of such a star as equal to one fifth of a second, we shall find that the sun advances through space, carrying with it the whole system of planets and comets, with a velocity about one fourth of the earth's annual motion in its orbit.

A great many questions here naturally suggest themselves. Is the sun's motion uniform and rectilinear, or is it moving slowly in an orbit about some center? Are the stars moving in straight lines, or in grand orbits? Have all the stars, including our sun, a common movement of rotation about some general center? This question has been examined by Professor Mädler of the Dorpat observatory, and he has attempted to assign the center round which the sun and stars revolve, which center he places in the group of the Pleiades.

If we assume that the orbit described by our sun about the central point is a circle, this central point must be found on the circumference of a great circle, whose pole is that point toward which the sun is moving, in right ascension $259\frac{1}{2}°$, declination $34\frac{1}{2}°$ north. This circle cuts the Milky Way in the constellation Perseus, and Argelander conjectured the central point to be here in Perseus. The most remarkable cluster of stars in this neighborhood is the Pleiades, and Mädler conjectured that here might be the central point. Accordingly he determined the proper motion of the eleven principal stars in this cluster by comparing the observations of

Bessel with those of Bradley and other astronomers. These motions exhibit considerable uniformity, and their direction is invariably toward the south.

Mädler next examined the 12 principal stars within five degrees of this group, and finds that 8 exhibit a decided southern motion, while in the other 4 the motion is too small to be decisive, but in no case is the motion toward the north. Among thirty stars, between 5 and 10 degrees distant from the Pleiades, Mädler finds that 20 are moving toward the south; while the motion of the remaining 10 is scarcely perceptible. Among 57 stars, between 10 and 15 degrees from the Pleiades, 16 are moving toward the south, while the motion of the remaining 41 is scarcely perceptible. Not one moves toward the north. Out of 66 stars, between 15 and 20 degrees from the Pleiades, 30 have a decided southern motion, and 36 are undecided. Thus, out of 176 stars, within 20 degrees of the Pleiades, we find 85 moving toward the south, and 91 whose motion is scarcely perceptible, but not a single case in which there is a considerable motion toward the north.

Mädler next examined all of Bradley's stars between 20 and 30 degrees from the Pleiades, of which the number is 175; of these, 78 exhibit a motion toward the south, 92 are uncertain, and 5 have a slow motion toward the north, amounting in the most rapid case to only seven seconds in a century. Such a result Professor Mädler considers a necessary consequence of his hypothesis. Since only small real motions are to be expected in the neighborhood of the central point, the motions, which are only apparent, and therefore contrary to the solar motion, must preponderate for all stars between the sun and the Pleiades.

The most rapid proper motions, according to this hypothesis, must be sought for near the great circle described about the Pleiades as a pole; and accordingly we find near this circle two of the most decided of all the proper motions hitherto discovered.

Professor Mädler accordingly infers that the central point of the starry heavens must be placed in the neighborhood of the Pleiades. This group is the nearest, the brightest, and the richest cluster in the whole heavens. Moreover, Alcyone is the optical center of this group, and he infers that this is the star which combines the strongest probability of being the true *central sun*.

Alcyone, known also as η Tauri, or 25 Tauri, is a double star of the third or fourth magnitude, in right ascension 3h. 38m., declination 23° 39′ north.

Assuming the parallax of 61 Cygni, as determined by Bessel, and that the sun and this star are moving with the same velocity about Alcyone, Mädler has computed that the distance of Alcyone is 34 millions of times that of the sun, requiring 537 years for its light to come to us, although moving at the rate of twelve millions of miles per minute. The periodic time of the sun about Alcyone is estimated at 18 millions of years; and the sum of the masses of all the stars within the sphere described about Alcyone as a center, with a radius equal

9

to the sun's distance, is 117 million times the mass of the sun.

The preceding conclusions are certainly wonderful ; but, unfortunately, they rest upon a very unsatisfactory basis. Professor Mädler has subjected the stars in the neighborhood of Alcyone to a very careful examination, and finds decisive indications of the sun's relative motion, as it had been previously established by the labors of Argelander and others. But beyond the neighborhood of the Pleiades, the number of stars examined is altogether too small to form the basis of so important conclusions. Sir John Herschel pronounces it almost inconceivable, that any general circulation of the stars can take place out of the plane of the Milky Way, while the Pleiades are situated 20 degrees out of this plane. It is not, however, presumptuous to expect that the problem which Mädler has propounded will one day be resolved. A careful determination of the proper motion of a considerable number of stars suitably situated in different parts of the heavens, could not fail to settle the question.

M. Struve has made a careful comparison of the places of the stars observed at Dorpat, with their places as determined by other astronomers, for the purpose of discovering their proper motion. The whole number of stars which formed the subject of research amounted to 1662, of which there were 734 single stars and 928 double stars. From these researches, M. Struve concludes that the angular motions of the different classes of stars are inversely proportional to their distances, from which he

infers that their linear motions are equal.   This he found
to be true for stars situated in every direction of the
heavens, by a comparison of the proper motions of the
stars in the different hours of right ascension.

The following table is founded on an examination of
the proper motions of the stars of different magnitudes as
deduced by M. Struve.

MEAN MOTION IN ONE HUNDRED YEARS.

Magnitude.		Isolated Stars.				Binary Stars.	
		R. A.	Dec.			R. A.	Dec.
1	. .	34″.2	29″.0	.	.	55″.5	47″.0
2	. .	18″.9	16″.1	.	.	30″.8	26″.1
3	. .	12″.4	10″.5	.	.	20″.1	17″.0
4	. .	8″.7	7″.4	.	.	14″.2	12″.0
5	. .	6″.3	5″.3	.	.	10″.2	8″.6
6	. .	3″.7	3″.1	.	.	6″.0	5″.1
7	. .	2″.2	1″.8	.	.	3″.5	3″.0
8	. .	1″.4	1″.2	.	.	2″.3	2 .0
9	. .	1″.0	0″.9	.	.	1′.7	1″.5

SECTION V.

THE last few years have been remarkable for the production of the largest telescope ever manufactured. Sir William Herschel constructed, with his own hands, telescopes of 20 and 40 feet focus, with which he made some of the most brilliant discoveries recorded in the history of astronomy. But quite recently, the Earl of Rosse has completed a telescope still more gigantic than the largest of Sir William Herschel. He had previously constructed a telescope of three feet aperture, which received the highest commendation from Dr. Robinson and Sir James South. In 1842, he commenced another of far superior dimensions, whose speculum was six feet in diameter, and weighed *three tons*. The materials of which it is composed are copper and tin, united in the proportion of fifteen parts of copper to seven of tin. The process of grinding was conducted under water, and the moving power employed was a steam engine of three horse power. The substance made use of to wear down the surface was emery and water, and it required six weeks to grind it to a fair surface.

The tube of the telescope is 56 feet long, and is made

of wood one inch thick, and hooped with iron. The
diameter of this tube is 7 feet. At 12 feet distance on
each side of the telescope, a wall is built, 72 feet long, 48
high on the outer side, and 56 on the inner, the walls being 24 feet distant from each other, and lying exactly in
the meridian. When directed to the south, the tube may
be lowered till it becomes almost horizontal; but when
pointed to the north, it only falls till it is parallel with
the earth's axis. Its lateral movements take place only
from wall to wall, and this commands a view for half an
hour on each side of the meridian; that is, the whole of
its motion from east to west is limited to 15 degrees. The
expense of this instrument was not less than *twelve thousand pounds*. It has a reflecting surface of 4071 square
inches, while that of Herschel's 40 feet telescope had only
1811 square inches.

In March, 1845, Sir James South made a trial of this
telescope, and gives the following account of his observations: "Never before in my life did I see such glorious
sidereal pictures as this instrument afforded us. The
most popularly known nebulæ observed were the ring
nebula in the Canes Venatici, which was resolved into
stars with a magnifying power of 548, and the 94th of
Messier, which is in the same constellation, and which
was resolved into a large globular cluster of stars, not
much unlike the well-known cluster in Hercules. On
subsequent nights, observations of other nebulæ, amounting to some thirty or more, removed most of these from
the list of nebulæ, where they had long figured, to that

of clusters; while some of these latter exhibited a sidereal picture in the telescope such as man before had never seen, and which, for its magnificence, baffles all description."

In the Philosophical Transactions for 1844, Lord Rosse has given some observations of nebulæ made with his three feet speculum, accompanied with drawings of the most remarkable objects. Among these is one which Sir John Herschel had figured as an oval resolvable nebula. Lord Rosse's telescope exhibits it with resolvable filaments singularly disposed, springing principally from its southern extremity, and not, as is usual in clusters, irregularly in all directions. It is studded with stars, mixed, however, with a nebulosity, probably consisting of stars too minute to be recognized. This has been called the *crab nebula*.

The Dumb Bell nebula, known everywhere by the drawing of Sir John Herschel, is seen to consist of innumerable stars mixed with nebulosity; and Lord Rosse remarks, that when we turn the eye from the telescope to the Milky Way the similarity is so striking, that it is impossible not to feel a conviction that the nebulosity in both proceeds from the same cause.

The annular nebula in Lyra shows filaments proceeding from the edge of the ring, and also several filaments partly filling up the interior of the ring. By the three feet speculum it was not resolved, but the filaments became conspicuous under increasing magnifying power, which circumstance is strikingly characteristic of a cluster.

The nebula in the Dog's Ear was formerly regarded

as a representation of our own Milky Way, and although unresolved, it was by common consent considered a mighty cluster. At the meeting of the British Association in 1845, Lord Rosse showed a sketch of its appearance as seen by aid of his six feet mirror. The former simple shape of this nebula is transformed into a *scroll*, apparently unwinding with numerous filaments, and a mottled appearance, which looks like the breaking up of a cluster.

The great nebula in Orion has been examined with every great telescope since the invention of that instrument, but until recently without the remotest aspect of a stellar constitution. During Sir John Herschel's residence at the Cape of Good Hope, he examined this nebula under the most favorable circumstances, when it was near the zenith—but still there was no trace of a star, only branches added without number, so as almost to obliterate the nebula's previous form. During the winter of 1844–5, Lord Rosse examined it with his three feet mirror with the utmost care, but without detecting the vestige of a star. In the winter of 1845–6, the six feet telescope was directed, for the first time, to this wonderful object, and in March, 1846, Lord Rosse made the following announcement: "I think I may safely say that there can be little if any doubt as to the resolvability of this nebula. We can plainly see that all about the trapezium is a mass of stars; the rest of the nebula also abounding with stars, and exhibiting the characteristics of resolvability strongly marked."

Mr. Bond, with the great telescope at Cambridge, has also seen this nebula partially resolved. To him the head of the nebula appears composed of several clusters of stars, the components being separately seen for a moment under favorable circumstances.

Mr. Lassell observed the nebula of Orion with his twenty feet equatorial at Malta under the most favorable circumstances. The following are extracts from his journal:

"1852, Dec. 6. Viewed the nebula with powers 219; 260 and 1018. With the latter power, a new phase was given to the nebula, which seemed like large masses of cotton wool packed one behind another; the edges pulled out so as to be very filmy.

"Dec. 8. I applied a power of 1018, with which there is no appearance of resolvability. The whole aspect is that of a number of masses of fleecy cloud, thin at the edges, and packed one behind another, appearing to be a deep stratum of successive layers of nebulous substance.

"Dec. 15. A few more stellar points, I believe, appear than I have mapped down in my Starfield diagram, and the stars contained in those diagrams are very much brighter. With power 1018, the wool-like masses appear as I have previously described them, and there is no disposition whatever in them to turn into stars."

The Great Nebula in Andromeda has also been carefully observed with the Cambridge telescope. The most conspicuous features were the sudden condensation of light at the center into an almost star-like nucleus; and the vast number of stars of every gradation of brilliancy scattered over its surface, which yet had the undefinable, but still convincing aspect of not being its components. It is estimated that above fifteen hundred stars are visible with the full aperture of the object-glass within the limits of the nebula. With high powers, minute stars are discovered on the borders of the nucleus, but it has thus far yielded no evidence of resolution. Minute descriptions, accompanied with accurate drawings of both these nebulæ, have been recently given by the Messrs. Bond in the Memoirs of the American Academy, Vol. III.

On the whole, it appears that the increase in the power of our telescopes has added to the number of the clusters at the expense of the nebulæ properly so called; still, as Lord Rosse has remarked, it would be very unsafe to conclude that such will always be the case, and thence to infer that all nebulosity is but the glare of stars too remote to be separated by the utmost power of our instruments. While Lord Rosse's telescope has shown certain nebulæ to contain an immense number of stars, it has also revealed to us new nebulous appendages of extreme faintness, which we must regard either as not composed of stars, or as composed of stars of very small absolute dimensions.

9*

CHAPTER IV.

SECTION I.

ASTRONOMICAL OBSERVATORIES IN THE UNITED STATES.

IT is but a few years since practical astronomy began to be cultivated in the United States in an efficient and systematic manner. Until recently, the instruments in our possession were but few and small, and the observations which were made, seldom extended beyond the notice of the time of a solar or lunar eclipse, or the measurement of a comet's distance from neighboring stars with a sextant.

The most important astronomical enterprise undertaken in this country, during the last century, was the observation of the transit of Venus in June, 1769. Upon the observations of this transit depended the more accurate determination of the sun's parallax; from which is deduced the distance of the earth from the sun, and thence the absolute distances of all the planets. Only three transits of Venus are known to have ever been seen by any human being. The first occurred in December, 1639, and was seen by but one individual, named Horrocks, who lived near Liverpool, England. The next transit occurred in June, 1761, and was carefully observed in different parts of the world, and important conclusions were drawn from it as to the sun's parallax. It was known, however, that its next recurrence, which was to take place in 1769, would be under more favorable circumstances, and several of the governments of Europe sent astronomers to various parts of the globe favorably situated for the observations. France sent an astronomer to California, England sent astronomers to Hudson's Bay, to Madras, and to the Island of Otaheite, in the South Sea. Several Russian observers were stationed at various points of Siberia and the Russian empire. The King of Denmark sent an astronomer to the North Cape, and the King of Sweden sent an observer to Finland.

The American Philosophical Society, in January, 1769, appointed a committee of thirteen to observe this rare phenomenon. The gentlemen thus appointed were distributed into three committees for the purpose of making observations at three different places: viz., in the city of Philadelphia; at Norriton, 17 miles northwest of Philadelphia; and the light-house, near Cape Henlopen, on Delaware Bay. Dr. Ewing had the principal direction of the observatory in the city, Mr. Rittenhouse at Norriton, and Mr. O. Biddle at Cape Henlopen. Some money was appropriated by the Philosophical Society toward defraying the expenses of the observations; but this being found insufficient, aid was solicited and obtained from the Assembly. Temporary observatories were erected, tolerably well adapted to the purposes for which they were designed. A reflecting telescope with a Dollond micrometer was purchased in London by Dr. Franklin, with the money voted by the Assembly; another of the same character was presented by Thomas Penn, of London; and other instruments were supplied in sufficient number. The observations at the three stations were all successful, and an account of them is given in the first volume of the Transactions of the American Philosophical Society.

For more than half a century after the transit of Venus, very little, if any, progress seemed to have been made toward the erection of a permanent observatory, or toward the procuring of large instruments such as modern astronomy requires. The first direct proposition for the establishment of an observatory was contained in Mr. Hassler's project for the survey of the coast, submitted to the Government through Mr. Gallatin, in the year 1807. The proposition met with no favor. The original law, authorizing the survey, passed without any provision on the subject, and the law of 1832 expressly prohibits such an establishment. The late John Quincy Adams, in his first annual message in 1825, strongly urged this subject upon the attention of Congress. After recommending the establishment of a National University, he said:

"Connected with the establishment of a university, or separate from it, might be undertaken the erection of an

astronomical observatory, with provision for the support
of an astronomer, to be in constant attendance of observation upon the phenomena of the heavens; and for the
periodical publication of his observations. It is with no
feeling of pride, as an American, that the remark may be
made that, on the comparatively small territorial surface
of Europe, there are existing upward of one hundred and
thirty of these light-houses of the skies; while throughout the whole American hemisphere there is not one. If
we reflect a moment upon the discoveries which, in the
last four centuries, have been made in the physical constitution of the universe by the means of these buildings,
and of observers stationed in them, shall we doubt of
their usefulness to every nation? And while scarcely a
year passes over our heads without bringing some new
astronomical discovery to light, which we must fain receive at second-hand from Europe, are we not cutting
ourselves off from the means of returning light for light,
while we have neither observatory nor observer upon our
half of the globe, and the earth revolves in perpetual
darkness to our unsearching eyes?"

This eloquent appeal from the chief magistrate of the
country, in behalf of the cause of science, was received
with a general torrent of ridicule; and the proposition to
establish a light-house in the skies became a common by-
word of reproach which has scarcely yet ceased to be
familiar to the lips of men who glory in their own shame.
So strong was this feeling that, in the year 1832, in re-
viving an act for the continuance of the survey of the

coast, Congress was careful to append the proviso, that
"*nothing in the act should be construed to authorize the
construction or maintenance of a permanent astronomical observatory.*"

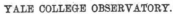

A donation made to Yale College by Mr. Sheldon Clark
is believed to have contributed somewhat toward that impulse which astronomy has recently received. In 1828,
Mr. Clark made a donation of twelve hundred dollars to
Yale College for the purchase of a telescope. The telescope was ordered from Dollond, of London; it arrived
in 1830, and was pronounced by the maker to be "per-

fect, and such an instrument as he was pleased to send as a specimen of his powers." This instrument has a focal length of 10 feet, and an aperture of 5 inches. The object-glass is almost perfectly achromatic. For objects that require a fine light, as the nebulæ and smaller stars, this instrument exhibits great superiority, and its defining power is equally good. It has a variety of eyeglasses, and a spider-line micrometer of the best construction.

The style of mounting of this telescope is not equal to its optical character. It has an altitude and azimuth movement without graduated circles, and is rolled about the room upon casters. The location of the instrument was peculiarly unfortunate. It was placed in the steeple of one of the college buildings, where the only view afforded of the heavens was through low windows which effectually concealed every object as soon as it attained an altitude of thirty degrees above the horizon. Under these circumstances, the telescope has proved less serviceable to science than might otherwise have been anticipated. On one occasion, however, circumstances gave this telescope considerable celebrity. The return of Halley's comet, in 1835, was anticipated with great interest. The most eminent astronomers of Europe had carefully computed the time of its appearance, and the results of their computations had been spread before the public in all the popular journals. All classes of the community were impatiently watching to learn the result of these predictions. The comet was first observed in this country by

Professors Olmsted and Loomis, with the Clark telescope, weeks before news arrived of its having been seen in Europe. This was the occasion of bringing prominently before the public the importance of having large telescopes, with all the instruments necessary for nice astronomical observations. It gave a new impulse to a plan which had already been conceived of establishing a permanent observatory at Cambridge, upon a liberal scale— a plan, however, which required the momentum of another and more splendid comet for its completion. It kindled anew the astronomical spirit of Philadelphia, and excited a desire for instruments superior to those which were then possessed. Indeed the importance of systematic astronomical observations was beginning to be somewhat generally felt, as well as the necessity of superior instruments for this purpose, and many embryo plans were formed for the establishment of astronomical observatories.

A transit instrument of five feet focal length and four inches aperture has recently been presented to Yale College by Mr. William Hillhouse, of New Haven; but for want of a suitable building for its reception, this instrument has not yet been mounted.

WILLIAMS COLLEGE OBSERVATORY.

The first attempt to found a regular astronomical observatory in this country was made in connection with Williams College, Massachusetts, by Professor Albert Hopkins. In 1836, Professor Hopkins erected a small

building, consisting of a center with two wings, the whole
being 48 feet in length by twenty in breadth. The central apartment is surmounted by a revolving dome 13
feet in diameter, and each wing has an opening through
the roof for meridian instruments. Under the dome was
placed a Herschelian telescope of 10 feet focus, mounted
equatorially. The circle for right ascension was a foot in
diameter; the declination semicircle was 30 inches in

WILLIAMS COLLEGE OBSERVATORY.

diameter. Both were made by Mr. Phelps, of Troy, New
York, and read to minutes. In the east wing has been
placed a transit instrument by Troughton, having a focal
length of 50 inches, and an aperture of three and a half
inches. In the same room is a compensation clock by
Molineux.

In 1852 an achromatic refracting telescope, having an
aperture of seven inches, and a focal length of nine and a

half feet, was presented to Williams College by the late
Amos Lawrence, Esq., of Boston. The optical part was
manufactured by Mr. Clark, of Boston, and the mounting
was furnished by Phelps. The instrument is mounted
equatorially, and has a clock movement. It will afford
some indication of the excellence of this instrument to
state that the sixth star in the trapezium of Orion has
been seen by it. This telescope has been mounted under
the dome of the observatory in place of the former reflecting telescope.

HUDSON OBSERVATORY, OHIO.

The next experiment for an observatory was made in
Ohio, in connection with the Western Reserve College.
Having been elected to the professorship of mathematics
and astronomy in this institution in the spring of 1836,
the writer was sent to Europe for the purchase of instruments and books, and returned in the autumn of 1837
with an equatorial telescope, a transit circle, and a clock.
During the next season a building was erected, which,
though quite moderate in dimensions, was well suited to
the accommodation of the instruments. The entire length
of the building is 37 feet, and its breadth 16 feet. The
transit room is 10 feet by 12 upon the inside, having a
sandstone pier in its center. The pier is entirely detached
from the building, and descends about six feet below the
surface of the earth. The transit commands an unob-
structed meridian from ninety degrees zenith distance on
the south, to eighty-nine on the north.

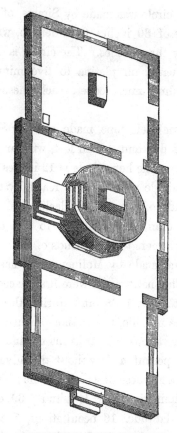

GROUND PLAN OF HUDSON OBSERVATORY.

The center room is occupied by the equatorial. It is 14 feet square on the inside, and is surmounted by a revolving dome of nine feet internal diameter. The equatorial pier descends six feet below the surface of the ground, and, like the transit pier, has a slope of one inch to the foot.

The transit circle was made by Simms, of London. It has a telescope of 30 inches focal length, with an aperture of nearly three inches. The circle is 18 inches in diameter, graduated on platina to five minutes; and it has three reading microscopes, each measuring single seconds.

The equatorial telescope, made also by Simms, has a focal length of five and a half feet, with an aperture of about 4 inches. The hour circle is 12 inches in diameter, graduated to single minutes, and reads by two verniers to single seconds of time. The declination circle is also 12 inches in diameter, graduated to ten minutes, and reads by two verniers to ten seconds of arc.

The clock was made by Molineux, and has a mercurial pendulum. The instruments were first placed in the observatory September, 1838, and during the whole time of his residence in Ohio, the author pursued a systematic course of observations, as far as his engagements in the college would permit, and without the advantage of an assistant. Among these observations may be mentioned, 260 moon culminations for longitude, 69 culminations of Polaris for latitude, 16 occultations, 5 comets, with sufficient accuracy to afford a determination of their orbits, besides a great variety of other objects, for regulating the clock, etc.

The moon culminations observed at Hudson have been compared with European observations, as far as corresponding ones were made, and the following is the result: On 119 nights the moon was observed both at Green-

wich and Hudson; on 107 nights it was observed both at
Edinburg and Hudson; on 95 nights at Cambridge (England) and Hudson; on 88 nights at Hamburg and Hudson;
and on 40 nights at Oxford and Hudson. The discussion
of all these observations, the results of which are published in Gould's Astronomical Journal, has furnished
the longitude of Hudson from Greenwich with a precision such as has been attained at but few other places
in the United States.

In the summer of 1849, the observatory at Hudson
was compared with that at Philadelphia by means of the
electric telegraph, numerous signals having been transmitted to and fro on four different nights, and the difference of longitude between these places has thus been
settled within a small fraction of a second. The accurate
determination of the geographical position of a single
such place in a new State, affords a standard of reference
by which a large surrounding territory is tolerably well
located through the medium of the local surveys.

PHILADELPHIA HIGH-SCHOOL OBSERVATORY.

The High-School observatory at Philadelphia was
erected at about the same time with that of Western
Reserve College, but the instruments were not received
until the autumn of 1840. In the year 1837, a committee was appointed by the Board of Controllers of
Public Schools on the subject of establishing a Central
High-School in Philadelphia. At one of the meetings
of the committee, Mr. George M. Justice proposed the

PHILADELPHIA HIGH-SCHOOL OBSERVATORY.

erection of an observatory to be attached to the school,
and that the use of astronomical instruments should be
taught as a regular branch of study. The committee
unanimously adopted the suggestion, and placed the
erection of the observatory and furnishing the instruments under the care of Mr. Justice. The Controllers
placed at the disposal of the committee the sum of $5000
to furnish the observatory. In accordance with the advice of Mr. Sears C. Walker, it was decided to order the instruments from Munich, in preference to London or Paris.

In the year 1838 a tower, about 45 feet high, was erected in the rear of the school building, and was insulated 10 feet below the surface of the earth. The brick walls were three feet thick at bottom, and two and a half feet thick at top, and the diameter of the tower was about 12 feet in the clear. It was surmounted by a dome 18 feet in diameter, weighing about two tons. The telescope rested on two marble slabs, each weighing about a thousand pounds, which were supported by two strong cast-iron beams that reached from the north to the south brick wall, and thus bound the two walls together.

The equatorial, by Merz and Mahler, of Munich, is of eight feet focal length, and six inches aperture, with clock-work movement. The hour circle is nine inches in diameter, reading to four seconds of time; the declination circle is 12 inches in diameter, reading to ten seconds of arc. This telescope is mounted like the celebrated telescope at Dorpat, and has a variety of powers to 480, with micrometers.

The meridian circle is by Ertel, of Munich, and was mounted on marble pillars resting on the south wall of the tower. The telescope has an object-glass of five feet focal length, four and a half inches aperture, and is so constructed that the object-glass and eye-glass may be made to change places. It has two circles, each graduated to read by the aid of four verniers to two seconds of arc. The clock is by Lukens, and has a mercurial pendulum. The cost of the several instruments was as follows: equatorial telescope, $2,200; meridian

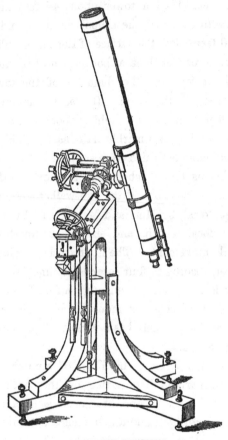

PHILADELPHIA EQUATORIAL.

circle, $1,200; clock, $300; comet seeker, $245; chronometer, $250.

The erection of this observatory formed an epoch in the history of American astronomy, in consequence of the introduction of a class of instruments superior to any which had been hitherto imported. It introduced the instruments of Munich fairly to the notice of the American public; and their superiority to the English telescopes was felt to be so decided, that almost every large instrument which has been since imported has been from the same makers. In the hands of Messrs. Walker and Kendall, this observatory became celebrated, not only in America, but also in Europe. It has furnished 436 moon culminations, about 120 occultations of stars, and several series of observations for latitude; together with numerous observations of comets, especially the great comet of 1843. This was also an important station in several of the earlier telegraph operations for longitude.

The ground occupied by the High School being needed for the accommodation of one of the railroads leading out of the city, the building and lot were sold in 1853, and a new lot was purchased on Broad-street, about 250 rods north from the former site. Here a new school building and observatory have been erected, and the instruments were set up in the autumn of 1854. The following is a description of the new observatory :

For the support of the instruments, two parallel piers of solid masonry, nearly at right angles to the plane of the meridian, each 16 feet wide and $2\frac{1}{2}$ feet thick, were erected in the central front part of the building. They are 18 feet apart, being separated by the main entrance-hall and principal stairway; the latter extending to the fourth story of the tower. The piers are inclosed and completely isolated, and extend from below the founda-

10

tions of the building to a height of 90 feet, terminating
about one foot above the ceiling of the fourth story of the
tower.   They are tied together at each floor by wooden
beams, and at the top by four cast-iron girders—a pair,
four and a half feet apart, being placed near each end.

The observing room is 24 feet square, and is covered
by a flat roof with the exception of the part occupied by
the equatorial.

The eastern pair of iron girders are framed together
midway between the piers, and upon this frame-work is,
constructed of bricks and cement, a prism, four feet
square at the base, reaching nearly to the floor, and from
this point a cylinder, three feet in diameter, rises three
feet above the floor.   This cylinder is incased in a drum
of boiler iron.   The marble stand of the telescope is im-
bedded to a depth of 18 inches in the cylinder, and rises
to a height of six feet and three quarters above it.   The
equatorial is covered by a hemispherical dome of 12 feet
in diameter, constructed upon the plan of that of Mr.
Campbell's observatory in New York, with revolving
table and steps attached.   To secure a firm support for
the dome, a frame-work of cast iron is constructed below
the floor, resting on the walls within the piers.   From
this frame-work, eight equidistant cast-iron columns extend to the ceiling, and upon these a substantial ring,
12 feet in diameter, composed of wood and iron, serves
as the foundation for the plate upon which the dome revolves.

The western end of the southern pier is extended to

within about one foot of the floor, and is capped by a
slab of marble, eight inches thick, upon which the piers
of the transit circle stand.   In the direction of the meridian of the transit is a clear opening, 26 inches wide, in
the roof and down the sides to within about two feet of
the floor.   On each side of the transit circle a flexible gas
tube hangs from the ceiling.   One light is used to illumi-
nate the wires of the transit, while the other is used in
reading the circle.

The clock is attached to the western wall, near the
transit instrument, and is lighted by gas.

Adjoining the observing room on the east, is a small
apartment which serves as a library and computing room.
This room is provided with a stove, for which reason the
two rooms do not communicate directly with each other,
but both open on the staircase leading from the fourth
story to the observatory.

WEST POINT OBSERVATORY.

The West Point observatory was erected about the
same time with that at Philadelphia.   In 1839, a large
building was erected for the accommodation of the library
and philosophical apparatus, with three towers for the reception of astronomical instruments.   The central tower
is surmounted by a traveling dome, 27 feet in diameter,
and about 17 feet high from the spring.   It is pierced by
five window-openings near the curb, and an observing
slit, two feet wide, extending from a point four feet above
the floor to nearly two feet on the opposite side of the

WEST POINT OBSERVATORY—NORTH FRONT.

zenith. The dome rests on six twenty-four pound can-non-balls, which turn between two cast-iron annular grooves.

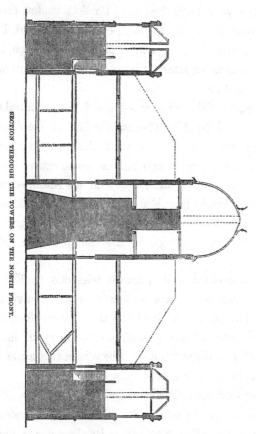

SECTION THROUGH THE TOWERS ON THE NORTH FRONT.

In the two flank towers, meridian observing slits are made, about 20 inches in the clear. These begin about two and a half feet from the floor, and extend through the roof, thus affording an uninterrupted view of the celestial meridian from the southern to the northern horizon.

In the year 1840, Professor Bartlett visited the principal observatories in England, Scotland, Ireland, France, Belgium, and Bavaria, and ordered three large instruments, viz., an equatorial telescope, a transit instrument, and a mural circle.

The equatorial, which was erected in the central tower, was mounted by Mr. Thomas Grubb, of Dublin. The telescope, made by Lerebours, of Paris, is a refractor of eight feet focal length, and six inches aperture. It has a position micrometer, furnished with an illuminating apparatus for bright lines and dark field. The telescope is moved by clock-work, so that the object under examination is easily kept in the center of the field of view.

In the east tower is a transit telescope, by Ertel and Son, of Munich. It has a clear aperture of five and a quarter inches, with a focal length of seven feet, and is supplied with all the appendages necessary to facilitate the making of observations. There is in this tower a fine sidereal clock, by Hardy.

In the west tower is a mural circle, by Simms, of London. It is cast in one entire piece of brass, instead of the old mode of frame-work. Its diameter is five feet, and the graduations are upon two bands, one of gold the other of palladium. The telescope has a clear aperture of four inches, and a focal length of five feet. It is provided with all the usual means of adjustment, together with a vertical collimating eye-piece, and an illuminating apparatus for dark field and bright lines. Professor Bartlett, the director of the observatory, has subjected this instrument to a severe trial, and finds the probable error in the measurement of an angle of 60° to be but 0″.22, exclusive of the error of reading. There is also a sidereal clock in the same tower.

A new refracting telescope, designed to take the place of the Grubb telescope in the central tower of this observatory, has just been completed by Mr. Henry Fitz. This telescope has a focal length of 14 feet, and an aperture of nine and three fourths inches. It has seven negative and six positive eye-pieces, the highest magnifying power being 1000. The circles are each 20 inches in diameter, the hour circle reading to two seconds of time, and the declination circle to 20 seconds of arc. The price of this telescope was $5000.

Professor Bartlett made a series of observations on the great comet of 1843, which are published in the "Transactions of the American Philosophical Society." He has also made numerous observations with the meridional instruments, which have not yet been published.

NATIONAL OBSERVATORY AT WASHINGTON.

Soon after the completion of the West Point observatory, the National observatory at Washington was commenced. The origin of this establishment may be traced to the wants of the naval service. In the year 1831, there was established at Washington a dépôt of charts and instruments for the use of the navy. A small transit instrument was erected in a small wooden building near the Capitol, and used for the rating of chronometers. This dépôt was for several years under the superintendence of Lieutenant (now Captain) Wilkes. When, in 1838, this officer took command of the exploring expedition, he recommended that a series of observations should be made

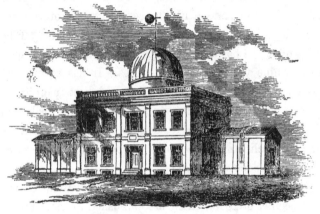

NATIONAL OBSERVATORY AT WASHINGTON.

in this country, during his absence, upon such celestial phenomena as might be available for the better determination of his longitudes, and their reference to some meridian at home. The government sanctioned the recommendation, and the observations were directed to be made at Dorchester by Mr. Bond, and at Washington by Lieutenant Gilliss. This series was continued until the return of the expedition, in 1842.

On his return from Europe, in 1840, Professor Bartlett made a report to the Engineer Department at Washington on the observatories of Europe. In this report he embodied the modern improvements in the construction of instruments, as well as the erection of observatories. He afterward prepared a plan and estimates for an observatory at Washington, for Mr. Poinsett, then Secretary of War.

In 1842 was passed an Act of Congress authorizing the erection of a dépôt of charts and instruments for the navy, the expense being limited to $25,000. Lieutenant Gilliss was instructed by the Secretary of the Navy to present a plan of a building, after consultation with the principal astronomers of the United States. The plan thus prepared was afterward submitted to the most eminent astronomers of Europe, and the model finally adopted embraced such improvements as they had recommended. The observatory consists of a central building of brick, with wings upon the east, west, and south sides. The central building is 50 feet square, two stories high, with a basement, and is surmounted by a revolving dome 23 feet in diameter, with an elevation of 18 feet from the floor. Directly under the dome is the great pier, whose diameter at the base is 15 feet, and tapers gradually to the top, upon which rests the great equatorial. The central building, except the dome, is employed exclusively for official purposes. The west wing is 21 by 26 feet, and 18 feet high, and is appropriated to the meridian transit instrument. The east wing is 48 by 21 feet, and

is divided into two rooms, in one of which are the mural
circle and the meridian circle, and in the other are the
chronometers. The wing on the south side is 21 by 40

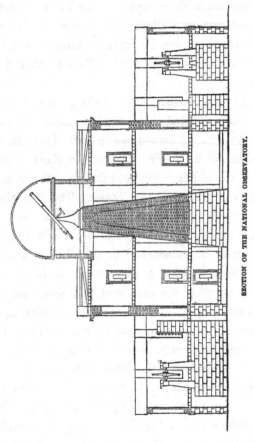

SECTION OF THE NATIONAL OBSERVATORY.

feet, in two apartments. In the first apartment is the
transit in the prime vertical, and in the second apartment
is the new refraction circle. In each of these wings is

a clock regulated to sidereal time. Immediately east of
the observatory is the dwelling of the superintendent.

The great refracting telescope was made by Merz and
Mahler, of Munich. The object-glass has a focal length
of 15 feet, and an aperture of nine and a half inches.
This telescope is equatorially mounted, and furnished
with clock-work. It has a repeating filar micrometer,
with eight eye-pieces, magnifying from 100 to 1000 times.
The cost of this telescope was $6000, its object-glass
alone being valued at $3,600.

The transit instrument has an object-glass with a clear
aperture of five and a half inches, and a focal length of
88 inches, furnished by Merz and Mahler, and the instrument was constructed by Ertel and Son, of Munich.
It is mounted upon large piers of granite, which rest
firmly on a foundation of stone, extending ten feet below
the surface of the ground. The cost of this instrument
was $1480; the object-glass alone cost $320.

The mural circle was made by Mr. William Simms, of
London. It is five feet in diameter, made of brass, and
cast in a single piece. It is divided into spaces of five
minutes each, upon a band of gold inlaid on the rim, or
perpendicular to the plane of the circle. Placed at equal
distances round the circle are six micrometer microscopes,
with an acute cross of wires at their foci, for reading
angles less than five minutes. Five revolutions of the
micrometer are designed to measure five minutes upon
the circle. The micrometer heads being divided into
sixty equal parts, each division represents one second of

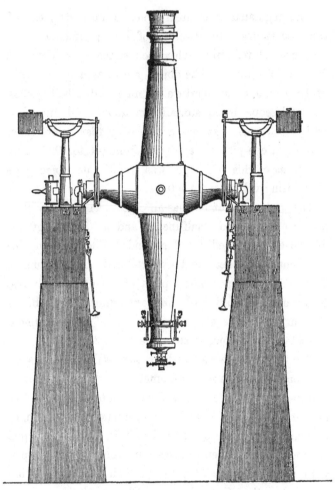

TRANSIT INSTRUMENT.

arc. The object-glass of the telescope has a clear aperture of four inches, and a focal length of five feet. At
its focus is a fixed diaphragm, containing seven vertical

wires and one horizontal wire. There is also another diaphragm, movable by a micrometer screw, and furnished with five horizontal and equidistant wires, by which the distance of any star from the fixed horizontal wire is measured as it passes through the field. The magnifying power ordinarily employed is 125. The cost of the mural circle was $3,550.

The meridian circle was made by Ertel and Son, of Munich. Its object-glass has an aperture of four and a half inches, with a focal distance of 58 inches. This instrument is provided with a circle 30 inches in diameter, divided into arcs of three minutes, and reads by four microscopes to single seconds. The clock in the east wing has a mercurial pendulum, and was made by Charles Frodsham.

The transit in the prime vertical was made by Pistor and Martins, of Berlin. The object-glass of the telescope has a clear aperture of five inches, with a focal length of 78 inches. The eye-tube carries a system of 2 horizontal and 15 vertical stationary wires, with one movable vertical wire. This instrument is mounted at one *end* of its axis and outside of its supports. It is reversed from one side to the other twice during every observation; and though it weighs upwards of 1000 pounds, so perfect is its system of counterpoises and the reversing apparatus, that a child can lift it from its supports, reverse and replace it in them, in less than one minute. The cost of this instrument was $1750. The clock in this wing has a gridiron pendulum, and was made by Charles Frodsham.

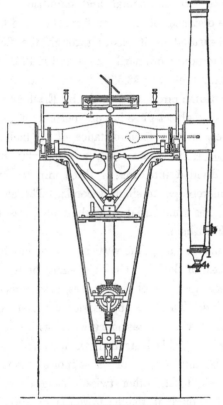

PRIME VERTICAL TRANSIT.

The refraction circle was made by Ertel and Son, from plans and drawings furnished by Lieutenant Maury. The telescope is eight and a half feet in length, with a clear aperture of seven inches. It is supported in the middle of the axis between two piers, and it has two circles of

four feet diameter, one on each end of the axis, divided
on gold into arcs of two minutes. Each circle is provided with six reading microscopes. The telescope has
two micrometers, one moving in azimuth, the other in
altitude.

The comet-seeker was made by Merz and Mahler.
It has an object-glass of about 4 inches in diameter,
and with a focal length of 32 inches, with which low
magnifying powers are used, that it may embrace a large
field, and collect the greatest possible quantity of light.
The cost of this instrument was $280.

In the fall of 1844, Lieutenant Maury was directed to
take charge of the new "Dépôt of Charts and Instruments." Lieutenant Maury commenced a regular and
systematic series of observations upon the sun and moon,
the planets, and a list of fundamental stars, comprising
those of the greatest magnitude and of the most favorable
positions, to be used as standard stars. He also undertook observations for a most extensive catalogue of stars.

This work contemplates a regular and systematic examination of every point of space in the heavens that is
visible at Washington, and of assigning position, color,
and magnitude to every star that the instruments are
capable of reaching. The following plan of sweeping
was adopted: The telescope of the mural circle is set
in altitude, and all the microscopes carefully read and
recorded, and the eye-piece is moved up and down so
as to cover a belt of about 50 minutes in declination.
The micrometer diaphragm is provided with a number

of parallel wires, the intervals of which have been carefully determined. Thus, in whatever part of the field a star appears, a micrometer wire is close at hand, and the star is bisected by the nearest wire, while the time at which it passes the several vertical wires is also noted. The number of the bisecting wire, and the reading of the micrometer being now entered, the observation is complete.

The observer thus keeps his eye at the telescope for hours at a time, and, under favorable circumstances, can observe with ease two or three hundred stars during the night. The transit instrument, by means of a micrometer moving in altitude, is converted into a difference of declination instrument, and occupies the adjoining belt above the mural, the two instruments being so set that ten minutes of declination are common to the field of both. The meridian circle in the same way occupies the belt below the mural. The next night the instruments change places, and go over the same ground, *i. e.*, the meridian circle covers the same belt to-night which on the former night was swept by the mural. The two lists are immediately compared, and should it appear that any of the stars have changed their position, the large equatorial is put in pursuit, to see whether they are fixed stars or not.

This great work contemplates the examination of every star down to the tenth magnitude in the entire visible heavens; and while it looks to the discovery of new planets and unknown stars, it also aims to detect the disappearance of any stars found in existing catalogues.

In December, 1849, the use of the electric clock was introduced at the Washington observatory, and the original method of observation was somewhat modified. In the west transit instrument was inserted a new diaphragm, having two systems of wires which included 75 spider-lines, viz., one system of vertical and one system of inclined wires. Each system is divided into groups of five wires each. By observing the transit of a star over a group of vertical wires, its right ascension is determined; and by observing its transit over a group of inclined wires, the difference of declination between this and other stars similarly observed may be computed. All these observations are recorded on a fillet of paper by means of an electric circuit, by simply pressing a key as the star is seen to pass each of the wires of the transit instrument.

Three quarto volumes of Washington observations have been published, viz., the observations for 1845, 1846, and 1847.

The volume for 1845 contains 550 pages, and furnishes a full description of the instruments employed, illustrated by numerous engravings. It also furnishes a large number of observations with the transit instrument, the mural circle, and the prime vertical transit. The volume for 1846 contains 676 pages, and besides observations, with the instruments used in 1845, furnishes also observations with the meridian circle and equatorial. All these observation are carefully reduced, and the places of the sun, moon and planets, are compared with their

predicted places as given in the *Nautical Almanac.* The
volume for 1847 contains 480 pages, and in its arrangement is similar to the preceding volume. These volumes
have placed our National observatory in the first rank
with the oldest and best institutions of the same kind in
Europe. But few observatories in Europe produce an
equal amount of work in a year, and in point of accuracy the observations compare well with those of
foreign institutions. It is expected that the volume for
1848 will be ready for the printer some time during the
present year, and that the succeeding volumes will follow with but little delay.

The observations for the star catalogue have not yet
been published. The number of stars already observed
is estimated at about 100,000, included between 16 and
45 degrees of south declination. The want of sufficient
force for reducing the observations has caused the delay
in their publication.

During the first two or three years of the operations
of the observatory, Lieutenant Maury devoted considerable time to observations, especially with the prime vertical transit and equatorial; but for several years his
time has been entirely engrossed by general superintendence, and he has been obliged to leave the observations
to his assistants. It has been customary to assign a lieutenant and a professor of mathematics to each meridional
instrument. Frequent changes have been made in the
lieutenants employed at the observatory; one set of
officers being ordered to sea, and another set being sent

to supply their place. But the professors of mathematics have continued with tolerable permanence, and have acquired a corresponding familiarity with the instruments, and the computations growing out of their use. The following notices contain the names of those who have contributed most to give character to the observatory.

Professor John H. C. Coffin graduated at Bowdoin College in 1834, and commenced duty at the observatory in January, 1845. He was immediately placed in charge of the mural circle, and devoted his time exclusively to that instrument until 1851, when his eyes began to suffer from the severe usage to which they had been subjected, and he made but few observations after that time. In 1853 he was detached from the observatory, and ordered to join the Naval Academy at Annapolis, where he is now employed in the department of instruction. Professor Coffin applied himself to his duties as an observer with indefatigable perseverance ; and the published volumes of the Washington observations sufficiently attest the amount and value of his labors.

Professor Joseph S. Hubbard graduated at Yale College in 1843, and commenced duty at the observatory in May, 1845. During the first year he had charge of observations with the transit instrument; in 1846 he was assigned to the meridian circle; in 1847 he was transferred to the equatorial; and since that time he has had charge of the prime vertical transit. Since 1850 he has been chiefly employed in the reduction of back observations. In 1852 and 1853 he undertook a series of observations on Alpha Lyræ for the determination of its parallax, but the unfavorable state of the weather during the month of December, when the maximum of parallax occurs, frustrated his expectations, and he was compelled to abandon the attempt. Professor Hubbard has contributed to the *Astronomical Journal* various papers which have secured him a high reputation among astronomers. Among these papers may be mentioned his researches on the great comet of 1843; on the orbit of Biela's comet; on the orbit of the planet Egeria, etc.

Professor Reuel Keith graduated at Middlebury College in 1845, and commenced duty at the observatory in August of the same year. He was immediately assigned to the meridian transit, and up to the present time has given his exclusive attention to that instrument. The published volumes of the observations show the fidelity with which he has discharged the duties assigned to him.

Professor Sears C. Walker graduated at Harvard University in 1825, and was attached to the observatory during the principal part of the year 1846. During this time he was mainly employed in computations respecting the planet Neptune; in the discussion of the latitude of the observatory; in determining the error of standard thermometers, etc.

Professor James Ferguson, for many years first assistant in the department of the Coast Survey, became connected with the observatory in 1848, and was immediately assigned to the equatorial, of which he has had sole charge to the present time. He has made numerous observations of comets, and of the small planets, and has had the good fortune to discover a new asteroid, Euphrosyne, being the only instance in which a primary planet has been first discovered by an American observer.

GEORGETOWN OBSERVATORY.

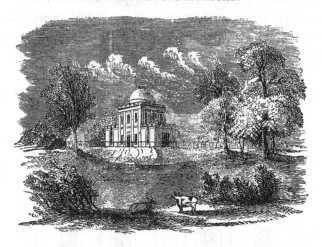

The erection of the Georgetown observatory was nearly cotemporaneous with that of the National observatory. In December, 1841, Rev. T. M. Jenkins offered a donation to the college at Georgetown for the purpose of building and furnishing an observatory ; and Rev. C. H. Stonestreet offered to supply an equatorial. In 1842 the donations were accepted. In the summer of 1843 the foundations of the building were laid, and it was finished in the spring of 1844.

The ground on which the observatory is built is 154 feet above the level of the Potomac river, from which it is distant about half a mile. The central part of the building is 30 feet square on the outside, with connecting wings both on the east and west sides, each of them being 27 feet by 15, making the entire length of the observatory 60 feet. The central part is surmounted by a rotary dome 20 feet in diameter, which works on cast-iron rollers, 8 inches in diameter. The opening in

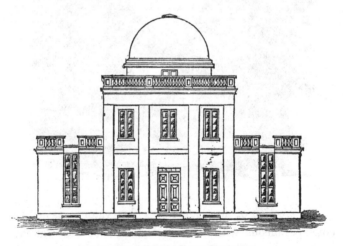

GEORGETOWN OBSERVATORY, SOUTH ELEVATION.

the dome is 2 feet wide, and is closed by 4 shutters. From the cellar, through all the floors of this part of the building, rises a pier of masonry 41 feet high. This pier is 11 feet square at the base, and 6 feet square at

the top, and upon it rests the equatorial telescope, made
by Simms, of London, which was received in 1849. The
object-glass has a focal length of 80 inches, and an
aperture of nearly 5 inches, with powers from 25 to 408.
The hour circle is 16 inches in diameter, and reads to
one second of time; the declination circle is 20 inches
in diameter, and reads to five seconds of arc. The instrument is supplied with clock-work, by which a celestial
object may be kept continually in the field of view. This
instrument cost $2000.

The east and west rooms, which contain the meridian
instruments, have meridian openings two feet wide
through the roofs, and down the north and south walls
to within two feet of the ground. In the west room is
mounted, on sand-stone piers, a transit instrument, made
by Ertel and Son, of Munich, which was received in
1844. The object-glass is four and a half inches clear
aperture and 76 inches focal length. It has a reversing
stand, by which the instrument can be reversed in a
minute and a half. This instrument cost $1180, be-
sides the expenses of transporting it from Munich. There
is also in this room a good sidereal clock by Molineux,
of London.

In the east room is mounted on two massive piers a
45 inch meridian circle, made, in 1845, by William
Simms, of London, with a telescope five feet long, and a
4 inch object-glass. The circle is graduated to five
minutes, and there are four micrometers fixed to the
eastern pier, reading to one second of arc. When the in-

strument is reversed, the readings are made by a second set of microscopes, which are attached to the western pier. In the eye-tube are seven fixed and one movable vertical wire, with one fixed and one movable horizontal wire. The lowest eye-piece is used for a collimating eye piece, by which the nadir point is determined by reflection from a vessel of mercury. The cost of this instrument was $2050. With this instrument there is a fine sidereal clock by Molineux, of London.

This observatory is under the direction of Rev. James Curley, who commenced a series of transit observations in 1846. During the autumn of the same year, he made some observations of circumpolar stars with the meridian circle, for determining the latitude of the observatory. During the year 1848, M. Sestini, of Rome, was added to this observatory, and it was expected that the celebrated comet hunter M. De Vico, of Rome, would be associated with Mr. Curley. But these expectations were suddenly disappointed by the death of M. De Vico, which took place at London, in November, 1848.

In 1852, Mr. Curley published a quarto volume of 216 pages, under the title of "Annals of the Astronomical Observatory of Georgetown College," giving a description of the observatory, as well as the transit instrument and meridian circle, and intimating that additional numbers of the "Annals" might be expected hereafter.

CINCINNATI OBSERVATORY.

The Cincinnati observatory owes its existence to the labors of Professor O. M. Mitchell. In the years 1841 and 1842, a society was organized in Cincinnati, called the Cincinnati Astronomical Society, the object of which was to furnish the city with an observatory. Eleven thousand dollars were subscribed in shares of twenty-five dollars each; and a site for the building was given by Nicholas Longworth, Esq. It consists of four acres of ground on one of the highest hills on the eastern side of

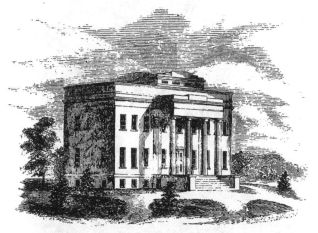

CINCINNATI OBSERVATORY.

the town. In June, 1842, Professor Mitchell visited Europe to purchase a telescope. At Munich, he found an object-glass of 12 inches aperture, which had been tested by Dr. Lamont, and pronounced one of the best

11

ever manufactured. This was subsequently ordered to
be mounted, and was purchased for $9437. The instru-
ment arrived in Cincinnati in February, 1845. In
November, 1843, the corner-stone of the observatory

CINCINNATI TELESCOPE.

was laid by the venerable John Quincy Adams. The
building is 80 feet long and 30 broad. Its front presents
a basement and two stories; while in the center the build-

ing rises three stories in height. The pier is built of stone, and is grouted from its foundation on the rock to the top. The equatorial room is 25 feet square, and is surmounted by a roof so arranged that it may be entirely removed during the time of observations.

The object-glass of the telescope has an aperture of 12 inches, and a focal length of 17 feet. The hour circle is 16 inches in diameter, and reads by two verniers to two seconds. The declination circle is 26 inches in diameter, and divided on silver to five minutes, reading by verniers to four seconds. The instrument has five common eye-pieces and nine micrometrical, with powers varying from 100 to 1400. It is furnished with clock-work, by which a star is kept steadily in the field of view of the telescope.

Through the liberality of Dr. Bache, the superintendent of the United States Coast Survey, this observatory has been furnished with a five feet transit instrument; and a new sidereal clock has recently been received.

Professor Mitchell has hitherto devoted much of his time to the measurement of Struve's double stars south of the equator. A number of interesting discoveries have been made in the course of this review. Stars which Struve marked as *oblong*, have been divided and measured; others marked *double*, have been again subdivided and found to be *triple;* while a comparison of the recent measures of distance and position with the measurements of Struve, has demonstrated the physical connection of the components of many of these stars.

For the last two or three years the energies of Professor Mitchell have been devoted almost exclusively to railroad engineering, and the observatory has consequently been neglected. We trust that he will soon free himself from such groveling occupations, and again direct the gaze of his powerful telescope to study the movements of distant worlds.

CAMBRIDGE OBSERVATORY.

The project of erecting an observatory in the neighborhood of Boston upon a scale corresponding with the importance and dignity of astronomy, had for a long period been the subject of conversation among the friends of science. This was a favorite scheme with John Q. Adams, Nathaniel Bowditch, and others, and various plans had been proposed for carrying it into execution; but it did not appear practicable to raise a sum of money sufficient to complete the plan upon the liberal scale which was desired. Something was needed to give a stronger impulse to the claims of practical astronomy. This impulse was given by the unexpected appearance of the splendid comet of 1843. In the month of March of that year, a comet with a long and brilliant train having made its appearance, the people of Boston naturally looked to the astronomers of Cambridge for information respecting its movements. The astronomers replied that they had no instruments adapted to nice cometary observations. This announcement, together with the knowledge of the existence of good instruments in other parts of the United States, aroused the general determination to supply the deficiency. Definite action was taken in March, 1843.

An informal meeting of a few individuals interested in the subject was held at the office of the American Insurance Company in Boston. The proceedings of this meeting were cordially seconded by the American Academy of Arts and Sciences, and a full meeting of merchants and other citizens of Boston was subsequently held at the hall of the Marine Society, to consider the expediency of procuring a telescope of the first class for astronomical observations. At this meeting the question was decided in the affirmative, and a subscription of twenty thousand dollars recommended to defray the expense. This amount was soon furnished. Mr. David Sears, of Boston, gave five thousand dollars for the erection of an observatory, besides five hundred dollars toward the telescope. Another gentleman of Boston gave one thousand dollars for the same object; eight other gentlemen of Boston and its vicinity gave five hundred dollars each; there were eighteen subscribers of two hundred dollars each; and thirty of one hundred

dollars each, besides many smaller sums.   The American
Academy of Arts and Sciences made a donation of three
thousand dollars; the Society for the Diffusion of Useful Knowledge gave one thousand dollars; the American,

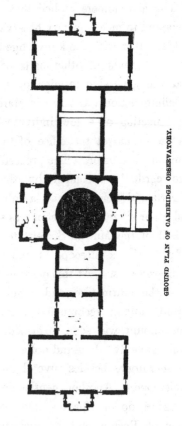

GROUND PLAN OF CAMBRIDGE OBSERVATORY.

Merchants', and National Insurance Companies and
Humane Society gave five hundred dollars each; two
other companies gave three hundred dollars each; one

gave two hundred and fifty, and another gave two hundred dollars.

The Corporation of Harvard University purchased an excellent site for the erection of an observatory. The ground comprises about six and a half acres. The position is elevated about 50 feet above the general plain on which are erected the buildings of the University; and it commands in every direction a clear horizon, without obstruction from trees, houses, smoke, or other causes. Upon this spot, which is known as Summer House Hill, the Sears Tower was erected for the accommodation of the large telescope, with wings for other instruments, and a house for the observer. The Sears Tower is a brick building 32 feet square, resting on a granite foundation. The corners of the towers are arched so as gradually to bring the interior into a circular form of 31 feet diameter, surmounted by a granite circle, on which is laid an iron rail hollowed in the middle to serve as a track for the iron balls on which the dome revolves. The dome has a diameter of 30 feet on the inside, with an opening five feet wide, extending beyond the zenith. The shutters to this opening are raised and closed by means of endless chains working in toothed wheels. To the lower edge of the dome is affixed a grooved iron rail similar to the one laid on the granite cap of the walls, and the dome rests on eight iron balls, which had been smoothly turned, and were placed at equal distances round the circle. Although this dome is estimated to weigh about fourteen tons, yet it can be turned through a whole revolution by a single individual, without any very great exertion, in thirty-five seconds.

The central pier for the support of the telescope is of granite, and is in the form of a frustum of a cone 22 feet in diameter at the base, and 10 feet at the top. It is 40 feet high, and rests on a wide foundation of grouting composed of hydraulic cement and coarse gravel, 26 feet below the natural surface of the ground, and is entirely detached from every other part of the building. Upon the top of the pier is laid a circular cap-stone, 10 feet in diameter, and 2 feet thick, on which stands, by three bearings, the granite block, 10 feet in height, to which the metallic bed-plate of the telescope is firmly attached by bolts and screws. Five hundred tons of granite were used in the construction of this pier.

Upon the east side of this tower is a small wing for the accommodation of the transit circle and clock; and on the north side is a similar wing, designed for a transit in the prime vertical. The house for the accommodation of the observer is connected with the east wing. The western wing is used for magnetic and meteorological observations. This wing was erected in the years 1850–51, making the entire length of the building 160 feet, and adds greatly to the architectural beauty of the observatory. In the small dome is placed the smaller equatorial, of 5 feet focal length, and 4 and 1-8 inch aperture, made by Merz, which is a remarkably fine instrument.

The "Grand Refractor" was made by Messrs. Merz and Mahler, of Munich, Bavaria. They bound themselves by contract to make two object-glasses of the clear aperture of 15 inches, to be at least equal to that furnished for the noble instrument now mounted at the Russian observatory at Pulkova. On being notified of the completion of these object-glasses, the agent of the University, Mr. Cranch, of London, accompanied by the instrument-maker, Mr. Simms, proceeded to Munich, and after careful trial and examination, made the required selection. The selected object-glass was received at Cambridge in December, 1846; the great tube and its equatorial mounting did not arrive until June, 1847. The object-glass of the telescope is 15 inches in diameter, and has 22 feet 6 inches focal length. Some of the eye-pieces are 6 inches long, making the entire length 23 feet. The telescope has eighteen different powers, ranging from 103 to 2000. The hour circle is 18 inches in diameter, divided on silver, and reading by two verniers to one second of time. The declination circle is 26 inches in diameter, divided on silver, and reads by four verniers to four seconds of arc. The movable portion of the telescope and machinery is estimated to weigh about three tons. It is, however, so well counterpoised in every position of the telescope, and the effects of friction are so far obviated by an ingenious arrangement of rollers and balance-weights, that the observer can direct the instrument to any part of the heavens by a slight pressure of the hand upon the ends of the balance

11*

rods.   A sidereal motion is given to the telescope by
clock-work, regulated by centrifugal balls, by which
means a celestial object may be kept constantly in
the field of view.   The cost of this instrument was
$19,842.

CAMBRIDGE EQUATORIAL.

The optical character of this instrument has given
entire satisfaction.   The components of the star Gamma
Coronæ, which Struve, with the Pulkova refractor, pro-

nounces *most difficult* to separate, being distant from each other less than half a second, are seen in the Cambridge telescope distinct and round, and the dark space between them is clearly defined. The components of Gamma Andromedæ, which are distant from each other less than half a second, are also separated with equal distinctness. The companion of Antares, estimated to be of the tenth magnitude, and which was discovered by Professor Mitchell with the Cincinnati refractor, is quite conspicuous with a power of 700. It was with this instrument Mr. Bond discovered the eighth satellite of Saturn, two days before it was discovered by Mr. Lassell, of Liverpool, with his Newtonian reflector of 21 inches aperture. He has also made satisfactory micrometric measurements of the satellite of Neptune, which is not known to have been done with any other instruments except Mr. Lassell's telescope and the Pulkova refractor. The minutest double stars in the neighborhood of the ring nebula of Lyra, mentioned by Lord Rosse as difficult objects with his 27 feet reflector, are seen in the Cambridge telescope. It has also partially resolved the great nebula in Orion, and shows a great number of stars within the limits of the nebula of Andromeda.

The transit circle was made by Simms, of London, and has been erected in the east wing. It has two circles, each of four feet diameter, graduated on silver to five minutes, and reads to single seconds by eight microscopes cemented to the granite piers, four microscopes belonging to each circle. The object-glass, furnished by Merz and

Mahler, has an aperture of four and one eighth inches, and a focal length of 65 inches. It has two different modes of illumination; one through the axis as usual, and the other at the eye-piece, showing bright wires on a dark field. Attached to the eye-piece are two microm-eters for measures both in altitude and in azimuth.

There is also belonging to the observatory a fine comet-seeker, by Merz and Mahler, having an aperture of four and a quarter inches. The wing on the north side of the tower is designed hereafter to receive a transit in the prime vertical; but this instrument has not yet been ordered.

During the summer of 1848, Mr. Bond being engaged with the United States Coast Survey in determining differences of longitude, turned his attention to the electro-magnetic method of recording astronomical observations. The apparatus adopted at this observatory consists of a Grove's battery, a circuit-breaking sidereal clock, and a "spring governor." These are connected by means of copper wires leading to all the principal instruments.

The *spring governor* is a machine devised to carry a cylinder with an equable rotary motion, so that it may make one entire revolution in one minute of sidereal time. A sheet of paper is wrapped round the cylinder, and on this paper the commencement of each second is recorded in exact coincidence with the beats of the clock. The observer at each telescope is furnished with a break-circuit key, by which means he is enabled to make a record of his observations on the paper covering the cylinder, among the second marks of the clock, in such a manner that the tenths of a second may be read off without difficulty.

The clock signals are also readily connected with the lines of the telegraph offices, so that in effect the beats of the Cambridge clock are as distinctly heard at the offices in Boston, Lowell, and elsewhere, as they are within a few feet of the clock. The time is thus given all along the telegraph lines, and this is found highly useful in regulating the starting of the railroad trains.

Mr. William C. Bond, and his son, George P. Bond, give their undivided attention to the objects of the observatory. For the first four or five years after receiving their grand refractor they gave their whole strength to that class of observations for which this instrument affords peculiar advantages, such as the following: observations of new planets; the satellites of Saturn, Uranus, and Neptune; double stars, especially such as have considerable proper motion; together with a general review of the most remarkable nebulæ. They have published in the Memoirs of the American Academy, a description of the great nebula in Orion, and that of Andromeda, accompanied with drawings of the most careful and elaborate execution.

The younger Bond for several years maintained a constant and systematic search for comets. With the comet-seeker he swept over the entire heavens at least once a month, and whenever he found any nebulous body with which he was not familiar, it was subjected to a special

examination.   He has thus been the independent discoverer of *eleven comets*, but unfortunately it subsequently
appeared that each of these, save one, had been previously discovered in Europe.   The comet of August 29,
1850, he discovered seven days in advance of the
European astronomers.   Two other comets he discovered
on the same night that they were seen in Europe, viz.,
those of June 5, 1845, and April 11, 1849.   Having
found this species of observation too severe a trial for his
eyes, he has for the last three or four years given up
comet-seeking altogether.

In April, 1852, the Messrs. Bond commenced a series
of observations which contemplate the formation of a most
extensive catalogue of stars down to the eleventh magnitude.   For this purpose they inserted in the focus of the
great equatorial a thin plate of mica, upon which were
ruled a large number of parallel and equidistant lines,
designed to measure differences of declination; and perpendicular to these they introduced three other parallel
lines designed for observations of right ascension.   The
telescope being fixed in the meridian, with the first set of
lines parallel to the horizon, if the times of passage of any
number of stars over the vertical lines be observed, we
shall have their differences of right ascension; and by observing near which of the horizontal lines they traverse
the field of the telescope, we shall obtain their differences
of declination.   The observations of right ascension are
all recorded by the electro-magnetic apparatus in the
manner already described.   The method pursued, there-

fore, by the observer, is first to fix the telescope firmly at any required altitude, and then applying his eye to the telescope, he observes each star in succession as it enters the field of view. One night's work embraces a zone ten minutes in breadth, and having a length corresponding to the number of hours for which the observations are continued.

The Messrs. Bond design to include in their series all stars to the *eleventh* magnitude, and as many of the *twelfth* as can be got without interfering with the determination of the brighter ones. Each zone is observed a second time as soon as possible, in order not to miss any chance of finding a planet which might happen to be in the way; but as the object is not planet-hunting, this is not allowed to interfere with the work, further than that missing stars are noted and looked after. It seldom happens that the disappearance does not prove to have originated in some mistake. At the re-observations, all the particulars of the first observation undergo a revision, the original notes being read and compared as the star passes the field. All the double stars, nebulæ, remarkable groups, vacancies, etc., are recorded. The comparison of the places of stars which have been previously observed at other observatories forms part of the reduction; and in this way they have detected some cases of considerable proper motion.

The Messrs. Bond have completed two entire zones (each twice observed) for the whole circuit of the heavens, from the equator to 20 minutes of north declination.

These have been completely reduced, and were published in 1855, under the title of "Annals of the Astronomical Observatory of Harvard College, vol. I., part 2." These two zones contain 5500 stars. The Messrs. Bond are now carrying on simultaneously, and have nearly completed, two zones from 20 to 40 minutes north declination.

The resources of this observatory have recently been very much increased by the munificence of Edward B. Phillips, a graduate of the University in the class of 1845. Mr. Phillips died in 1848, and bequeathed to the observatory $100,000 as a perpetual fund, the interest to be applied annually to the payment of the salaries of the observers, or for instruments, or a library for the use of the observatory, at the discretion of the corporation of the college, who are made the trustees of the fund. This sum was paid to the college by Mr. Phillips's executors in September, 1849.

SHARON OBSERVATORY.

Sharon observatory is a private establishment belonging to the late Mr. John Jackson, situated near Darby, about seven miles west of Philadelphia. It was erected in 1845, is 17 feet square on the outside, and rises to the height of 34 feet. At five feet from the top, two strong beams are placed across the building, and support a circular platform of five feet diameter, on which the equatorial stands. The tower is surmounted by a conical dome, which rests on four iron balls revolving in a circular railway. The opening in the dome is 18 inches wide, and has three doors which slide over each other, and are moved by cords passing over pulleys.

SHARON OBSERVATORY.

The equatorial was made by Merz and Son, of Munich. It was ordered in 1842, and arrived in 1846. The object-glass has a clear aperture of six and a third inches, and its focal length is nearly nine feet. It has a micrometer with a large number of eye-pieces magnifying from 85 to 456 times. The hour circle is 9 inches in diameter, and the declination circle is 13 inches. Clock-work is attached to the polar axis, giving to the telescope a uniform motion, and keeping a star apparently at rest in the field of view. The entire expense of this instrument was $1833.

This observatory is also furnished with a meridian

circle made by Young, of Philadelphia ; the object glass
having been procured from Merz and Son, of Munich.
It is mounted on a marble column, resting on solid
masonry, in the south wall of the tower. The object-
glass is three and a quarter inches in diameter, and has a
focal length of four feet. One end of the axis carries a
circle 20 inches in diameter, which is graduated to four
minutes, and reads by four verniers to three seconds.
The price of this instrument was $800. A sidereal clock
is supported by a marble column near the meridian circle.
It has a mercurial pendulum, and was made by Gropengiesser, of Philadelphia. The entire cost of this observatory, with the instruments, was $4000. The equatorial is
employed in the observation of eclipses, occultations, and
other celestial phenomena.

TUSCALOOSA OBSERVATORY, ALABAMA.

The Tuscaloosa observatory was erected in the year
1843, and has been furnished with instruments for observation of a superior order. It is situated upon an
elevation distant a few hundred yards from the University buildings, in a south-west direction. It is built of
brick, and is 55 feet in length by 22 in breadth in the
center. The central apartment is 22 feet square, and is
surmounted by a revolving dome of 18 feet internal
diameter, under which is mounted the equatorial telescope made by Simms, of London. This instrument was
received in the spring of 1849. Its object-glass has a
clear aperture of eight inches, and a focal length of 12

feet. It rests upon a pier of masonry which rises 11 feet above the floor of the room. The hour circle has a diameter of 18 inches, with verniers which read to one second of time. The declination circle has a diameter of 30 inches, with verniers which read to five seconds of arc. This instrument is moved by clock-work, and has a variety of magnifying powers from 44 to 1640; also a filar micrometer, and a double-image micrometer. A movable platform runs around the room at the height of eight feet from the floor, and gives easy access to the object-end of the telescope and the graduated circles. This telescope cost £800 sterling. In the same room is a clock with mercurial compensation by Molineux, of London, which is attached to a solid pier unconnected with the floor or walls.

TUSCALOOSA OBSERVATORY.

The west wing is 16 feet square, and is occupied by a transit circle made in 1840 by Simms, of London. Its telescope has a focal length of five feet, and an object-glass of four inches clear aperture. Its axis carries a circle of three feet diameter, connected with it by twelve stout conical radii. The circle is graduated upon silver to five minutes, and reads by four microscopes to single seconds. The whole instrument is mounted upon massive cast-iron pillars, which rest upon masonry, and rise five feet above the floor. An opening 18 inches wide is cut through the roof, and extends down the north and south walls to within 30 inches of the floor. In the same room is an excellent sidereal clock by Dent, of London.

The east wing is fitted up for an office, with fire-place, cases for books, etc.

Beside the fixed instruments here mentioned, the observatory possesses a portable transit instrument, an achromatic refractor of 7 feet focal length, a reflecting circle of 10 inches diameter, standard barometer, etc.

The University has a separate building for a magnetic observatory, and possesses a declination instrument and a dipping needle, both made by Gambey, of Paris.

MR. RUTHERFORD'S OBSERVATORY.

There is a private observatory, erected in the upper part of the city of New York, corner of Second Avenue and Eleventh-street, belonging to Lewis M. Rutherford, Esq. It is furnished with a refracting telescope, made by Henry Fitz, of New York. The aperture of the object-glass is nine inches, and its focal length nine and a half feet. It was mounted equatorially, with clock-work, like

the Dorpat telescope, by Messrs. Gregg and Rupp, of
New York. The hour circle is eighteen inches in
diameter, and the declination circle eleven inches. The
telescope has four eye-pieces, the highest magnifying 600
times. The price of this instrument, including clock-
work and micrometer, was $2,200. The telescope rests
upon a brick column, surmounted by a revolving dome
of twelve feet diameter.

Connected with this observatory is a small building
containing a transit instrument by Simms, belonging to
Columbia College. The telescope has an aperture of
nearly three inches, and a focal length of four feet. It is
mounted upon two stone columns, resting upon a solid
foundation of sandstone. An opening in the roof affords
a view of about 160 degrees of the meridian. In a small
wing of this building is a stone column, upon which is
placed an altitude and azimuth instrument by Simms,
also belonging to Columbia College. The horizontal and
vertical circles are each fifteen inches in diameter,
graduated to five minutes, and reading by two microscopes to one second of arc. The telescope has an aperture of two inches, and a focal length of twenty-four
inches.

During the summer of 1848, this observatory was
employed by the Coast Survey as a station for determining the difference of longitude between Cambridge and
New York by means of the electric telegraph.

FRIENDS' OBSERVATORY, PHILADELPHIA.

This observatory is situated in the city of Philadelphia, about 400 feet east of Independence Hall. It was built in 1846, and has a revolving dome fifteen feet in diameter. In the center is the stand for the equatorial, which rests on the walls of the building, unconnected with the floor of the observatory. The principal instrument is a refracting telescope of five inches aperture, and seven feet focal length, made by Henry Fitz, of New York, and mounted equatorially after the manner of Fraunhofer, by William J. Young, of Philadelphia. The hour circle is nine inches in diameter, and the declination circle twelve inches.

A twenty-inch transit instrument is permanently placed on a pier built on the wall of the building. A clock, with a mercurial pendulum, made by J. L. Gropengiesser, of Philadelphia, is placed on a pier adjoining the transit instrument. The observatory has also a portable refracting telescope of three inches aperture, and forty-two inches focal length, made by Chevalier, of Paris, and a comet-seeker of three inches aperture, mounted on a tripod, made by Henry Fitz, of New York.

Since the establishment of this observatory, occultations, eclipses, etc., have been regularly observed by the director, Mr. Miers Fisher Longstreth, who has recently distinguished himself by his successful labors in the construction of new Lunar Tables.

AMHERST COLLEGE OBSERVATORY.

A small building for astronomical observations was erected in 1847 in connection with Amherst College, Massachusetts. This building consists of an octagonal tower fifty feet high, and seventeen feet in diameter, with a revolving dome, and a central pedestal for supporting a telescope. On the east side of the tower is attached a transit room, 13 by 15 feet, with a sliding roof. Here is mounted a transit circle, made by Gambey, of Paris, the telescope having a focal length of about three feet, and an aperture of two and a half inches. The circle is fifteen inches in diameter, graduated to five minutes, and is furnished with four verniers reading to three seconds. The clock was made by Breguet, and has a gridiron pendulum.

A large telescope, manufactured by Mr. Alvan Clark, of Cambridge, Massachusetts, has recently been received, and mounted under the dome. The aperture of this telescope is seven and a quarter inches, and its focal length eight and a half feet. A block of stone, weighing about 1800 pounds, standing on the top of the brick pier, supports the instrument. The top of the stone is level, and a frame of cast-iron, clamped to the stone, sustains the polar axis. The clock, which gives the equatorial movement, is let into the top of the stone, and is regulated by a pendulum, while the motion is rendered continuous and nearly uniform by means of an elegant contrivance of Professor W. C. Bond, called the spring governor. With this telescope Mr. Clark discovered two new double stars, in one of which the distance of the components is but three-tenths of a second. This instrument cost $1,800, and was presented to Amherst College by Hon. Rufus Bulloch, of Royalston.

CHARLESTON OBSERVATORY, SOUTH CAROLINA.

This observatory was built by Professor Lewis R. Gibbes, in his own garden in Charleston. It is a wooden building of 10 by 15 feet, with a sliding roof. It contains a five feet transit instrument, by Troughton, solidly mounted on a pier of red sandstone, and commands the meridian to within ten degrees of the horizon on either side. There is also a five feet telescope, by Adams, of London, mounted equatorially, and provided with a position micrometer, by Troughton and Simms, and two chronometers, by Hutton, of London, one sidereal and the other solar. The instruments all belong to the Coast Survey, except the equatorial telescope.

Observations were commenced by Professor Gibbes in April, 1848, on moon culminations and occultations for the use of the Coast Survey, and have been continued to the present time. In February, 1850, this observatory was connected by the telegraph wires with the observatory at Seaton Station, Washington City, for difference of longitude by the electro-chronographic method. In March, 1851, it was connected in a similar manner with a temporary observatory in Savannah, Georgia, and a successful series of observations made for the difference of longitude of the two stations. During the winter and spring of 1852, another telegraph comparison for longitude was made between Charleston and Washington, which proved entirely successful.

DARTMOUTH COLLEGE OBSERVATORY.

The founding of Dartmouth College observatory was due chiefly to the munificence of the late George C. Shattuck, M.D., LL.D., of Boston, who furnished the means for the erection of the building, the purchase of the meridian circle, the comet-seeker, a chronometer, etc., together with a large part of the books belonging to the library. The site of the building is near the summit of an isolated hill, about 50 rods north-east of the college green, and elevated 70 feet above it, commanding a charming prospect up and down the Valley of the Connecticut. The principal south meridian mark is situated on the bare summit of a hill two miles distant, at an elevation of less than two degrees above the level of the transit circle.

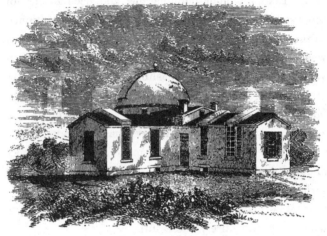

DARTMOUTH COLLEGE OBSERVATORY.

The observatory is of brick, with double walls fifteen inches thick, inclosing a six-inch space of air between them ; and, as the cornice and the partitions are also of brick, and the roof covered with tin, it is nearly fireproof. The building consists of a central two-story rotunda, 20 feet in diameter, with three one-story wings— one on the east, measuring 35 by 16 feet, and the other two on the north and south, each 20 by 16 feet. The foundations of the walls and of the piers all rest upon the solid rock—sienitic gneiss—at depths varying from zero to 14 feet below the underpinning. The lower cir cular room in the rotunda, 17 feet in diameter, is intended for a library, and communicates directly with the observer's rooms in the north and south wings. The square brick equatorial pier, four feet in diameter, rises through the middle of the room without touching the flooring or ceiling, and is capped with a cylindrical granite block nearly 6 feet in diameter, and 15 inches thick. The pier contains a recess, having double glazed doors, and a space of dead air on all sides, for the recep- on of a clock, to be connected with an electro-chronograph. The pier is also surrounded by the book-shelves, which are entirely detached from it.

The observer's rooms are in the north and south wings, and are each 21 by 14 feet, and have each a slit in the roof two feet wide, the north one having a corresponding pier for prime vertical observations.

The transit room, 18 by 14 feet, in the east wing, has two slits, the eastern one having a corresponding pair of piers for the meridian circle, and the other a single pier for the use of portable instruments. The entrance hall, 14 by 11 feet, in the center of the building, opens directly into all the rooms except the library.

The foundation for the dome of the equatorial room consists of circular segments of planks, firmly secured together, and bolted to the walls below. Upon this, and directly over the middle of the inner wall, are screwed twelve segments of a circular cast-iron rail, three inches wide, with a circular channel three eighths of an inch deep. Between this and a corresponding rail attached to the base ring of the dome, are placed the six cannon-balls, each six inches in diameter, on which the dome revolves. The dome itself is a complete hemisphere, 18 feet in diameter. The opening for the telescope is two and a half feet wide, and extends one foot beyond the apex of the dome. The opening for the

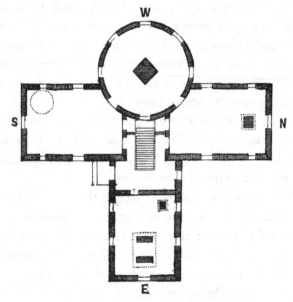

GROUND PLAN OF DARTMOUTH COLLEGE OBSERVATORY.

telescope is closed by three shutters; the upper one, about 10 feet long, moving on casters over an iron railway on the outside of the dome. The two lower ones, about two and a half feet each, rise and fall by means of weights in the manner of common window-sashes. The entire dome is estimated to weigh about 2800 pounds, and the average force necessary to preserve a uniform velocity of rotation is found, by experimenting with spring steelyards, to be about six pounds.

The equatorial telescope was made by Merz and Sons, of Munich; and has a clear aperture of six inches, with a focal length of eight and a half feet. It has seven negative eye-pieces, magnifying from 36 to 600 times; a single lens magnifying 940 times; a prismatic reflector, to which these pieces are all adapted; and a terrestrial eye-piece, magnifying 80 times. It has two micrometers, one a ring micrometer, the other a filar position micrometer, having 11 equidistant fixed lines perpendicular to the two movable ones, and may be illuminated in either a dark or bright field at pleasure. It has five positive eye-pieces, magnifying from 128 to 480 times. The hour circle is nine and a half inches in diameter, and the declination circle is thirteen inches in diameter. The telescope is moved in right ascension by an adjustable centrifugal clock.

The meridian circle, by Simms, of London, is 30 inches in diameter, divided on silver to spaces of five minutes, with two reading microscopes fixed to the piers, and a third placed at a small distance from one of the others, for examining and determining the errors of graduation. The telescope has a clear aperture of four inches, and a focal length of five feet. It has four positive eye-pieces, besides one for collimating by means of a mercurial trough. It is furnished with a declination micrometer and seven equidistant meridian lines. The wires can be

illuminated in a dark field, or the reverse, at the option
of the observer.   The granite piers for this instrument
are of the T form, gradually diminishing upward to
within 10 inches of the top.   They are 30 inches apart,
and rise 5 feet above the floor.   They rest on a granite
foundation, which is based on the solid rock below.

The comet-seeker, by Merz and Sons, has an aperture
of about 4 inches, and a focal length of 32 inches.   It is
mounted on an equatorial stand, and has 4 eye-pieces,
magnifying from 12 to 40 times.

The sidereal clock is furnished with Mahler's compensation pendulum, and runs a month without winding.
The library contains about 500 volumes of the most
valuable books pertaining to theoretical and practical
astronomy.

The cost of the building, exclusive of the lot and
its embellishments, was about $4,500.   The cost of the
equatorial telescope, together with the sidereal clock,
was about $2,300.   The comet-seeker cost $180; and
the meridian circle cost £275 in London.

MR. VAN ARSDALE'S OBSERVATORY, NEWARK, NEW JERSEY.

This observatory was built in the spring of 1850.   It
is a circular wooden building 13 feet in diameter, resting
on a brick foundation, underlaid with stone to the
depth of two feet.   The circular brick foundation is
covered with wood-work three inches in thickness, and
upon this are placed the wheels upon which the whole
building turns.   The wheels are eight in number, of cast-

ıron, and grooved. The circular iron rail attached to
the lower part of the building traverses the upper
surface of the wheels one foot from the floor. The roof
is of tin, and conical; the aperture is 16 inches wide,
and is covered by a door which opens in a single piece.
The pier upon which the telescope rests is built of stone
faced with brick, and rises two feet above the floor.

The telescope is a refractor, made by Henry Fitz.
The object-glass has an aperture of six and a half inches,
and a focal length of eight feet; and it was mounted
on an equatorial stand by Phelps and Gurley. The
declination circle has a diameter of ten inches, and
reads by verniers to one minute of arc; the hour circle
has a diameter of seven inches, and reads by verniers
to six seconds of time. The telescope is driven by
clock-work, and cost $1,125.

The object-glass of this telescope has given entire
satisfaction. Saturn is seen with a division in the outer
ring, the division being somewhat eccentrical and nearer
the outer edge. Occasionally there has been seen a sub-
division of the inner bright ring at the ansæ, near the
inner edge. The interior dark ring is also seen.

The planet Mars is well defined in a good atmosphere with a white circumference, both at the equatorial
and the polar regions, the interior space being varied
with dark shades.

The comet-seeker was also made by Henry Fitz.
The object-glass has an aperture of four inches, and is
fitted to a section of tube which screws into the finder

of the telescope. This mode has the advantage of a ready reference to the larger instrument, but is somewhat inconvenient from the interference of the stand in observing the space beneath the pole and near the meridian.

Three comets have been independently discovered by Mr. Van Arsdale, viz., one November 25, 1853, another June 24, 1854, and a third September 13, 1854. The first of these was not seen in Europe until December 2, a week after it was discovered by Mr. Van Arsdale; the second was discovered in Europe June 4; the third was discovered at Berlin, September 12. These discoveries were made while searching for comets, although the search was not thorough, as the view near the horizon is obstructed by buildings, and the examination was seldom continued to a late hour.

SHELBY COLLEGE OBSERVATORY.

A very superior telescope was ordered in 1848 from the establishment of Merz and Mahler, of Munich, for the use of Shelby College, Shelbyville, Kentucky. It has an aperture of seven and a half inches, and a focal length of ten feet. The mounting is admirably executed, and differs in some respects from any other in this country. It is furnished with a filar and an annular micrometer. The filar micrometer has six positive eye-pieces, with powers from 100 to 570. There are also 5 negative eye-pieces, magnifying from 102 to 550 times. The hour circle is 10 inches in diameter, reading to 4 seconds of time; the declination circle is 15 inches in diameter, reading to 10 seconds of arc. This instrument was received in November, 1850, and cost $3,500.

This telescope was mounted on the top of the college building, about 50 feet from the ground, under a revolving dome 18 feet in diameter, and was supported by a heavy cast-iron tripod, whose legs consist of hollow iron columns, each weighing about 600 pounds, passing through three floors of the building, and resting on solid masonry below the lower floor. The dome revolves on cannon-balls, and is turned by wheels gearing into a circle of cogs on the wall below the dome. The time is furnished by a box chronometer.

The cost of the telescope and observatory was defrayed partly by subscription; but the greater part of the expense was borne by the president of the institution, the Rev. William J. Walker.

BUFFALO OBSERVATORY.

This observatory is the property of Mr. William S. Van Duzee. It was erected in the summer of 1851, and the instruments mounted in the following winter. The building is 24 by 25 feet square, having two piers, one for the transit, and the other for the equatorial instrument, resting on a foundation of solid masonry, extending twelve feet below the surface of the earth, and surrounded by a wall two feet distant from this foundation, so that no motion may be felt by the passing of carriages.

The central pier is built of brick, and is 8 feet square through the first story, which is 15 feet high, but through the second story is only 6 feet square. This pier is surmounted by a traveling dome, 17 feet in diameter, and is pierced by an opening 2 feet wide, extending from a point 4 feet above the floor to 8 inches on the opposite side of the zenith. The dome rests on twelve eighteen-pound balls, which turn between two cast-iron annular grooves.

The transit pier is five feet square, and extends three feet into the second story. The meridian observing slits are made two feet in the clear, and afford an uninterrupted view of the meridian from the northern to the southern point of the horizon. Both piers are covered with marble slabs, three inches thick.

The equatorial telescope, which is erected under the dome, was made by Henry Fitz, of New York. It is a refractor of 11 feet focus and 9 inches aperture; and has a position micrometer furnished with an illuminating apparatus. The hour circle is 12 inches in diameter, and the declination circle 15 inches, each reading by two verniers. The telescope is moved by clock-work, so that the object under examination may be kept steadily in the center of the field. This telescope is furnished with ten eye-pieces, the highest magnifying 800 times.

The transit instrument is in the second story, has a clear aperture of 2 inches, and a focal length of 33 inches. Near the transit is the clock with a mercurial pendulum.

MR. CAMPBELL'S OBSERVATORY, NEW YORK.

This observatory is built on the top of Mr. Campbell's dwelling-house in New York city, in Sixteenth-street, near the Fifth Avenue, and was completed in 1852. The house is 30 feet wide and four stories high. The hall is 10 feet wide, and the partition wall which separates the hall from the rest of the house is of brick, and extends to the roof. This and the adjoining gable were raised so as to make another story over that part of the house, and a room was thus obtained 10 feet wide and 35 feet long. Twelve feet at one end are appropriated for the dome and telescope. The observatory is furred off so as to make an octagon of 12 feet in diameter; and at the height of 5 feet, the octagon is changed into a circle to support the wooden curb which constitutes the bed-plate of the dome. Upon the bed-plate is placed a circular rail, 12 feet in diameter on the inside, 3 inches wide, and having a raised bead upon the upper surface.

The dome is 12 feet in diameter, the base being a counterpart of the curb, which constitutes the bed-plate. The aperture for the telescope is 15 inches wide, and extends a little beyond the zenith. It is closed with a sliding door, which is made so that it may slip over the zenith to the opposite side of the dome, and this motion is effected by turning a crank. The dome revolves upon seven wheels of four inches diameter, in which grooves are turned to correspond with the bead on the iron rail. The shaft of one of these wheels is made long enough to receive a pinion at one end and a handle at the other. The pinion fits a rack which encircles the rail; while the handle and also the operator move round with the dome, which is accomplished by the peculiar construction of the observer's seat. This is a small flight of stairs, having an angle of elevation suited to the sweep of the eye-piece, so that each step makes a convenient seat. The frame is of wrought iron; the string pieces are secured to the base of the dome, and the bottom of the stairs rests upon two wheels, in which grooves are made to fit a circular iron rail which is secured to the floor. A circular table is substituted for the ordinary reading steps. It is supported upon legs, with rollers, and is secured at each end to the bottom steps of the observer's seat so as to revolve with it.

Three stout beams rest upon the brick walls across the center of the octagon, making a support for the pedestal of the telescope. This pedestal is a drum of boiler iron, three feet in diameter, and the same in height. It is lined with brick like a well, and covered with a smooth, round flag-stone, projecting an inch over the iron. The mahogany frame of the telescope, having four feet with adjusting screws, stands upon this stone.

The telescope is an achromatic refractor of eight inches aperture, and ten and a half feet focal length, made by Henry Fitz. It has six negative eye-pieces, magnifying from 60 to 480 times. The stand is made of cast iron, excepting the circles, which are of brass; and a clock is attached to the telescope for keeping the object in the field of view. The thread which fits the tangent screw is cut in the edge of the ring, detached from the hour circle, and pressed against it by the elasticity of a thin brass plate, which is secured by screws so as to give the requisite friction. This arrangement permits the telescope to be moved in any direction, even while connected with the clock, and obviates the necessity of clamping and unclamping.

The solar eclipse of May 26, 1854, was carefully observed by this telescope, and a series of daguerreotypes (28 in number) were taken at intervals of about five minutes, showing the progress of the eclipse in a novel and beautiful manner. These images, which are two inches in diameter, are extremely well defined, and serve as permanent registers of the transient phenomena of the eclipse.

OBSERVATORY OF MICHIGAN UNIVERSITY.

Soon after the Rev. Dr. Tappan was elected Chancellor of Michigan University, in 1852, he conceived the idea of establishing an astronomical observatory in connection with the university. During the succeeding winter the claims of this object were laid before the citizens of Detroit, and they responded to the appeal in a prompt and liberal manner. The sum of ten thousand dollars was almost immediately pledged for this object, and an additional sum was promised if it should be found necessary.

The foundations of the building were laid in the summer of 1853, and the observatory was completed in the

autumn of 1854. In February, 1853, a large equatorial
telescope was ordered from Henry Fitz, of New York;
and during the summer of the same year, a meridian
circle was ordered from Berlin. The main building for
the observatory is 32 feet square, and it is surmounted by
a hemispherical dome, 23 feet in diameter. The dome

OBSERVATORY OF MICHIGAN UNIVERSITY.

has an opening 32 inches wide, extending from near its
base to the zenith, and the opening is closed by a single
curved shutter, which slides in a groove, so that it may
be made to travel over to the opposite side of the dome.
A pier, whose foundations are laid ten feet beneath the
surface of the earth, rises through the center of the building to the height of 20 feet from the ground, and upon
this rests the great equatorial. On the east and west

sides of the central building are wings 28 feet by 19; the east wing being appropriated to the meridian circle, and the west wing being appropriated for the apartments of the astronomer. The cost of the entire building has been about $7000.

The meridian circle was made by Pistor and Martins, of Berlin. It has a telescope of eight feet focal length, and has two graduated circles of three feet diameter, each of which is read by four microscopes. There are also two small circles, with levels near the eye-piece for setting the instrument to the proper altitude. The eight microscopes are illuminated by two stationary lamps, which stand on the brass columns which support the counterpoises. The illumination of the wires can be so regulated as to present either a bright field with dark lines, or a dark field with bright lines. Two collimators are mounted on the north and south of the meridian circle, and there are conveniences for observing stars, as well as the nadir point, by reflection from a basin of mercury. The circle, with the collimators, cost $3300, and this sum was contributed by H. N. Walker, Esq., of Detroit. The clock was made by M. Tiede, of Berlin. It has a pendulum with a steel and zinc compensation, and cost $300.

The great equatorial telescope was made by Mr. Fitz, of New York. The object-glass has a clear aperture of twelve and a half inches, and a focal length of seventeen feet. It has seven negative and six positive eye-pieces, the highest magnifying power being 1200. The circles are each twenty inches in diameter, the hour circle reading to two seconds of time, and the declination circle to twenty seconds of arc. It is also furnished with clockwork and micrometer. The crown-glass was obtained from Bontemps, of Birmingham, England, and the flint-glass was obtained from Paris. The price of this telescope was $6000, and its performance has proved entirely satisfactory. The entire cost of the building and instruments was about $17,000.

Dr. F. Brünnow, an astronomer well known by his observations while connected with the observatory of Berlin, and also by his Treatise on Spherical Astronomy, has been appointed director of this observatory, and has already entered upon his duties. We congratulate the University of Michigan in having secured the services of so skillful and learned an astronomer; and we anticipate that the career of Dr. Brünnow will shed luster upon his adopted country, while he contributes to the promotion of that science to which his life has been devoted.

CLOVERDEN OBSERVATORY, CAMBRIDGE, MASSACHUSETTS.

The great telescope belonging to Shelby College was temporarily loaned to Professor Joseph Winlock, and was removed to Cambridge, Massachusetts, where temporary accommodations were provided for it, and this establishment is known by the name of Cloverden Observatory. This consists of a circular wooden building, with a revolving dome, fifteen feet in diameter, and a small adjoining room for other instruments. The dome is very light, consisting of a wooden frame covered with

tarpaulin, and revolves on wooden balls with so little resistance, that it is easily pushed round with the hand
without wheels or levers. In the adjoining apartment
is a four-feet transit and several chronometers. The
standard chronometer was made by Kessels.

Numerous observations on comets, and some of the
newly discovered planets, have been made with this
telescope by Dr. B. A. Gould and Professor Joseph
Winlock, some of which have been published in
"Gould's Astronomical Journal." The great telescope
has recently been returned to Shelby College.

DUDLEY OBSERVATORY, AT ALBANY.

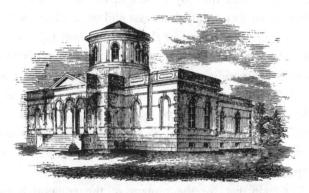

About three years since, it was proposed to establish
at Albany a university, comprehending a series of practical, professional, and scientific schools. As a part of
this enterprise, it was resolved to establish an astronomical observatory, the charge of which Professor
O. M. Mitchell had already signified his willingness to

accept. General Steven Van Rensselaer kindly offered a donation of several acres of land near the northern limit of the city, affording an excellent site for the contemplated building. Mrs. Blandina Dudley, a lady distinguished for her wealth and liberal spirit, gave toward the building the sum of $13,000. Contributions were received from several gentlemen of the city, increasing this sum to $25,000. The building was commenced in the spring of 1853, upon a plan designed by Professor Mitchell, and the late Sears C. Walker, and was erected under the supervision of Professor Perkins.

The ground-plan of the building is in the form of a cross, 84 feet in front by 72 feet in depth. The central room is 28 feet square; the east and west wings, which are designed for the use of the meridian instruments, are each 23 feet square, and provided with the usual openings in the meridian. The north wing is 40 feet square, divided into a library room, two computing rooms, and other small rooms for the magnetic apparatus for recording the observations. The equatorial room, which is in the second story, is of a circular form, 28 feet in diameter, the tower revolving upon iron balls.

The main pier for the support of the equatorial was commenced six feet below the bottom of the cellar, with its base, 15 feet square, resting on a bed of concrete and rubble 16 inches in thickness. The size of the pier was gradually reduced to 10 feet square at the level of the cellar, and thus continued upward without further variation. The whole is built in the most substantial manner of large stone, well bedded. The piers in the transit rooms are six feet by eight, and each room is furnished with clock piers of similar construction. The walls are of brick; but the basement, portico, cornices, etc., are of dressed freestone. The library and computing rooms are warmed by heated air from a basement furnace. These rooms and the staircase are all accessible from the vestibule. A bust of the late Charles Dudley, executed by E. D. Palmer, an artist of Albany, is to be placed near the principal entrance.

The Dudley observatory was incorporated in 1853, and its management is vested in a board of trustees. In the summer of 1855, Mrs. Dudley offered to furnish the requisite means for procuring a heliometer of the largest dimensions. The construction of this instrument has been entrusted to Mr. Charles A. Spencer, who proposes to make a telescope with an object-glass of 10 inches aperture. The price of this instrument is to be $14,500. A gentleman of Albany contributed $5000 for a meridian circle, and this instrument has been ordered from Pistor and Martins, of Berlin. Another gentleman of Albany contributed $1000 for an astronomical clock. The transit instrument will be furnished by the U. S. Coast Survey, and will cost $1,500. It is expected that the observatory will soon commence active operations under the direction of Dr. B. A. Gould.

HAMILTON COLLEGE OBSERVATORY.

In 1852 and 1853, Professor Charles Avery raised $15,000 by subscription, mostly in small sums, for the purpose of procuring a large telescope of American manufacture. It was then determined to erect on the grounds of Hamilton College a building suitable for an astronomical observatory, and to furnish it with first-class

HAMILTON COLLEGE OBSERVATORY.

instruments. In the spring of 1854, Mr. A. J. Lathrop, of Utica, presented plans of a building, which were approved, and the building was erected the same year.

The observatory consists of a central building, with wings on the east and west sides. The main building is 27 feet square, and two stories high, surmounted by a

tower, 20 feet in diameter, which revolves on eight cast-
iron balls upon an iron track.   The wings are each
18 feet square. The east wing is designed for a computing room; the west wing is designed for a transit
room, and it commands a fine meridian.   The transit
pier is built of limestone.   It descends six feet below the

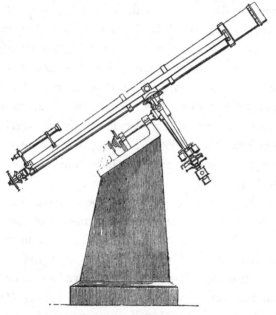

HAMILTON COLLEGE TELESCOPE.

surface of the ground, and is eight feet by five at the
top.   In the center of the main building is the large
pier, built also of limestone.   It descends six feet below
the surface of the ground, is ten feet square at the base,

and seven feet square at the top, and is entirely detached from the building. On the pier rests a granite block, fourteen feet high, to which the great refractor is to be attached. The observatory building cost $5000.

A refracting telescope was ordered from Messrs. Spencer and Eaton, of Canastota, New York, and has recently been completed. The object-glass has a diameter of 13¼ inches, with a focal length of 16 feet. The hour circle, divided on silver, has a diameter of 15 inches, and reads by two verniers to two seconds of time. The declination circle has a diameter of 26 inches, and reads to four seconds of arc. The filar micrometer has eye-pieces made of double achromatics, instead of the usual Ramsden form, giving magnifying powers from 100 to 1000. The negative eye-pieces give magnifying powers from 80 to 1500 times. The usual sidereal motion is given by clock-work.

The focal length of this telescope is about four feet less than is usual in the Munich instruments of the same aperture. The telescope is thus rendered less unwieldy, and in the opinion of Mr. Spencer, its optical performance is not impaired. The flint and crown discs for this instrument were procured through the agents, Messrs. Cook, Beckel and Co., New York, from Joseph Bader, of Kohlgrub, and have been found to be remarkably exempt from striæ. The object-glass has been subjected to partial trial with very satisfactory results. The price of this telescope was $10,000. The transit instrument and clock have not yet been ordered.

From the preceding sketch it must be apparent that within a few years, rapid progress has been made toward supplying our country with the means of making astronomical observations. We now have instruments which permit us to engage in astronomical researches upon a footing of equality with the oldest establishments of Europe, while the number of observers and the taste for astronomical studies has kept pace with the increase of our instruments. The importance of astronomical observations is beginning to be generally appreciated; but for the benefit of those whose attention has not been particularly turned to this subject, a few suggestions are added.

An astronomical observatory may be made useful in almost any locality, but it may be rendered especially valuable in the neighborhood of a large commercial emporium. The following may be enumerated among these advantages:

I. It furnishes an accurate determination of time, which is a matter of importance to almost every citizen, but more especially to certain classes of the community. Every vessel which puts to sea, carries with it one or more chronometers, and the chronometer is almost exclusively relied upon to furnish the longitude of the vessel from day to day. An error of a few seconds in the chronometer causes a corresponding error in the longitude deduced, and such errors have been the occasion of the loss of numerous vessels. It is, therefore, a matter of vital importance that the error and rate of all chronometers which are carried to sea should be determined with the utmost precision. In every commercial city there are private establishments where this duty is regularly attended to. A public observatory does not necessarily interfere with these private establishments, but affords the means of rendering them more accurate and efficient. How this may be accomplished will presently be shown.

An exact knowledge of time is also of vital importance to the conductors of all railroad trains. A small error in a conductor's watch has repeatedly been the occasion of the collision of railroad trains, and the consequent destruction of human life. Many of the railroad companies in this country incur annually considerable expense to provide all the conductors with correct time.

An accurate knowledge of time is important to all business men, but especially to banking and other houses where business is entirely suspended at a fixed hour of the day. A small mistake in the time might occasion not only serious disappointment, but also pecuniary loss.

An astronomical observatory is furnished with clocks of the best construction, and with transit instruments for determining daily the error of these clocks. The observatory, therefore, can furnish time with all the precision which can be desired; but to render this knowledge conveniently accessible to the public, requires some peculiar arrangements. The following is the arrangement for this purpose which existed for many years at Greenwich observatory: On one of the turrets of the observatory is erected a mast, upon which slides a ball, five feet in diameter, consisting of a frame of wood covered with leather. A little before one o'clock every day the ball is slid up to the top of the mast, where it is held by ingeniously contrived machinery. Precisely at one o'clock an assistant, who is specially charged with this duty, presses a spring, and the ball instantly descends. By this means all persons in sight of the ball are enabled daily to test the accuracy of their clocks. At the Washington observatory a ball of smaller size than that at Greenwich is elevated every day on a flag-staff, and is lowered at the precise instant of twelve o'clock.

Within the last two years the arrangements at Greenwich for furnishing the public with an accurate knowledge of the time have been very much improved. A normal clock is furnished with a small apparatus, by means of which, whenever its error is determined by observations, its indications can be rendered perfectly correct. This clock keeps in motion a sympathetic galvanic clock at the entrance-gate of the observatory, and also a clock at the terminus of the South-eastern Railway. It sends galvanic signals every day along all the principal railways diverging from London; it drops the Greenwich ball and the ball on the offices of the Electric Telegraph Company in the Strand. The Lords Commissioners of the Admiralty have also erected a time signal-ball at Deal, for the use of the shipping in the Downs, which is dropped every day by a galvanic current from the Royal observatory.

Similar arrangements might be adopted in every large commercial city. If, for example, there was a public observatory in the neighborhood of New York, a clock at the observatory might be made every day, by means of an electric current, to drop a time-ball on the Merchants' Exchange or the City Hall, another on Brooklyn Heights, another on Staten Island, and another at Sandy Hook; as well as at any other point where public convenience might require. It might also maintain in motion a sympathetic galvanic clock at the City Hall, at the Custom-house, at the Exchange, and at every railway station in the city—a clock which should never differ by an appreciable quantity from perfectly accurate time. Such a system would contribute not a little to the security of commerce and the punctuality of business.*

II. A second advantage to be derived from an astronomical observatory is that, by extending our knowledge of the heavenly bodies, it directly contributes to the security of commerce. The prosperity of commerce depends entirely upon the safety with which the ocean can be navigated, and this depends upon the accuracy with which a ship's place can be determined from day to day. Had it not been for the labors of modern astronomers in their observatories, vessels would still, as in ancient times, creep timidly along the coast, afraid to venture out of sight of land; or if they were compelled to venture

* Since the preceding was written, the officers of the Dudley observatory at Albany have offered to furnish time for the city of New York, substantially in the manner above suggested.

into the open ocean, they would be exposed to imminent danger in approaching land, not knowing how far distant the port might be. The loss of time resulting from pursuing this timid course, and the numerous disasters which could not be avoided, would more than double the expense of maintaining our foreign commerce. Astronomers, by their accurate determinations of the places of the sun, the moon, and the stars, have given prosperity to commerce and boundless wealth to our commercial cities. But there is still much for astronomers to do The places of the heavenly bodies even at present are not known with all the precision which is desired. Great errors in determining a ship's place are now of rare occurrence; but small errors frequently lead to disastrous consequences, and it is therefore important to reduce the errors to the least possible amount.

III. An astronomical observatory, well equipped, becomes a center of influence which is felt on all the educational establishments of the country, even those of the humblest grade. It is impossible to maintain common schools in a state of efficiency without institutions of a higher grade, corresponding to our academies, which shall furnish teachers for the elementary schools, and also afford encouragement to ambitious scholars, who wish to extend their studies beyond the elementary branches. The academies can not be maintained in a flourishing condition without institutions of a higher grade, corresponding to our colleges, where teachers are educated for the academies, and where ambitious students from the academies may extend still further the range of their studies. College professors, in their turn, are in danger of settling down into mere retailers of other men's ideas, without aspiring to add any thing to the stock of human knowledge, unless they are surrounded by institutions whose leading object is the increase of knowledge. An astronomical observatory, therefore, is a center of genial influence, which directly or indirectly imparts life and efficiency to all the subordinate institutions of education. It is also a place where men of business may acquire new ideas of the wonders of the material universe; where men, whose days are spent in toiling for the acquisition of wealth, may learn that there are mines of intellectual riches more inexhaustible than the mines of California. Men who from morning to night are engaged in the duties of an arduous profession, or in the labors of the counting-house or exchange, often feel the need of recreation when the hours of business are over. What mode of recreation is more rational—what is better fitted to inspire the mind with noble sentiments—than to direct the thoughts to the wonders of the material universe, to the vastness of the visible creation, as exhibited to the eye of an astronomer with the assistance of the telescope?

SECTION II.

ASTRONOMICAL EXPEDITION TO CHILI DURING THE YEARS
1849—1852.

In the year 1847, Dr. C. L. Gerling, a distinguished
mathematician of the Marburgh University, suggested
the importance of a new determination of the sun's
parallax by observations upon Venus at and near her
stationary periods. The determination of the dimensions
of the solar system rests entirely upon the assumed
value of the sun's parallax. The value now generally
received, viz., 8".57, rests upon the observations of the
transit of Venus in 1769. Transits of Venus over the
sun's disc afford the best method of determining this
parallax; but these phenomena are of very rare occurrence, there being not a single transit visible in any
part of the world from 1769 to 1874. Now, although
the observations of the transit of 1769 are believed to
have afforded a very accurate value of the sun's parallax,
yet it is much to be regretted that the results obtained
by combining the observations at different stations two
and two, differ among themselves by an entire second.
It is therefore very desirable that this result should be
verified by independent methods. Such methods are

found in simultaneous observations of either Venus or
Mars from two remote points of the globe. If an astronomer in a high northern latitude observes the position of one of these bodies when upon his meridian, and
another astronomer in a high southern latitude does
the same, a comparison of these two observations will
give the parallax of the planet, from which we can
compute its distance from the earth. The most favorable time for observing Mars is when it is nearest the
earth; that is, at its opposition; and the most favorable
time for observing Venus is when it is stationary, or
near its inferior conjunction. The two places of observation must be in opposite hemispheres, as remote as
possible from each other, and it is desirable that they
should both be under the same meridian..

Lieutenant Gilliss, of the United States Navy, proposed an expedition to Chili, for the purpose of making
observations upon Dr. Gerling's plan in connection with
the National observatory at Washington. The American Philosophical Society, and the American Academy
of Arts and Sciences commended this plan to the favorable action of Congress; and in August, 1848, Congress
authorized the fitting out of the expedition, under the
direction of the Secretary of the Navy. Lieutenant
Gilliss, to whom the plan of the expedition was mainly
due, was appointed to take charge of the expedition;
and passed midshipman (now lieutenant) Archibald
MacRea, and Henry C. Hunter, were appointed as assistants. Suitable wooden buildings, to serve as an ob-

servatory in Chili, were prepared in Washington, and shipped to Valparaiso in the summer of 1849.

The principal astronomical instruments furnished for the expedition were two telescopes, equatorially mounted, a meridian circle, a clock, and three chronometers.

The larger telescope was an eight and a half feet refractor, having an object-glass by Fitz, of New York, that afforded a clear aperture of six inches and a half. It was fitted with clock-work by Wm. Young, of Philadelphia, and by him provided with a micrometer adapted both for differential measurements, and for measurements of position and distance.

The other telescope was a five feet achromatic, by Fraunhofer. It was also equatorially mounted, and fitted with a micrometer by Young, of Philadelphia.

The meridian circle was by Pistor and Martins, of Berlin. The object-glass of the telescope had a clear aperture of four and one-third inches, with a focal length of six feet. The circles were thirty-six inches in diameter, both divided to $2'$, and read by two micrometer microscopes, each to a half second.

The series of astronomical observations, especially contemplated, consisted of differential measurements during portions of the years 1849, 1850, 1851, and 1852, upon Venus and Mars, with stars conveniently situated in the neighborhood of their paths. The observations of Venus chiefly depended upon, were observations near the inferior conjunctions of 1850 and 1852. The principal observations of Mars were near the times of opposition of that planet in 1849 and 1852. To facilitate the observations, and to secure concert of action, so that observers in every part of the world might use the same stars of comparison, Lieutenant Gilliss prepared an ephemeris of these planets and suitable stars of comparison during the critical periods.

The expedition set sail in 1849, and arrived safely at Valparaiso. The observatory was located at Santiago, the capital of Chili, where observations were commenced in December, 1849. During that season the weather was exceedingly favorable for observations. Of the fifty-two pre-appointed nights remaining of the first series of observations on Mars, there were only four when no observations could be made; and within these forty-eight nights, 1400 observations of the planet were accumulated. During the second series on Mars, comprising 93 days, between December 16, 1851, and March 15, 1852, about 2000 differential measures were made on seventy-eight nights, and meridian observations on eighty nights. Between October 19, 1850, and February 10, 1851, differential measures of the planet Venus and comparing stars, were made on fifty-one nights; and there were seventy-three meridian observations, at which time its diameter also was measured. Owing to its very frequent tremulous motion in the evening twilight, the differential measures, when approaching its eastern stationary terms, were often found difficult, and rarely afforded much satisfaction. But in the morning twilight the atmosphere was tranquil, and generally so clear that measures could be continued long after day-light, if the comparing star was as bright as the seventh magnitude. During the second period of Venus in 1852, it was possible to make differential measures only on nine evenings prior to the conjunction, and on eighteen mornings subsequent to it. There was not a single occasion when the measures were wholly satisfactory.

As the preceding observations occupied only a part of the time spent in Chili, Lieutenant Gilliss devoted a portion of the intermediate intervals to constructing a catalogue of stars between the south pole and 65° of south declination. Beginning within 5° of the south pole, a systematic sweep of the heavens was commenced in zones or belts 24' wide. Working steadily toward the zenith on successive nights, until compelled to return below again to connect in right ascension, the place of every celestial body that passed across the field of the telescope, to stars of the tenth magnitude, was carefully noted down. The space immediately surrounding the south pole was swept in one belt of 5° by moving the circle. Above the polar belt, forty-eight other belts were observed, making in all 24° 12' of declination, within which were obtained 33,600 observations of some 23,000 stars; more than 20,000 of them never previously tabulated.

While astronomical observations received the principal attention, magnetism and meteorology were not neglected. The meteorological instruments were observed tri-hourly during nearly three years, and one day of each month

13*

was devoted to hourly observations. The magnetical observations occupied nearly four hours of two and sometimes three persons on three days of each month, when all the elements necessary for determining the direction and the total force of the earth's magnetism were carefully observed.

A minute record was preserved of all the earthquakes experienced during the continuance of Lieutenant Gilliss in Chili. The number of shocks recorded was 125 in 34 months, being an average of about one each week. A seismometer was erected, but it was not sufficiently sensitive, and generally made no record of the agitations of the earth's crust.

Congress liberally ordered the publication of the results of this expedition, directing that the whole work should be well printed on good paper of quarto size, and well bound. The results will be comprised in seven volumes, of which the first two have already been published. Vol. I. contains a full account of Chili, its geography, climate, earthquakes, government, social condition, mineral and agricultural resources, commerce, etc., with a narrative of the general progress of the expedition. Vol. II. contains Lieutenant MacRea's narrative of his magnetical expedition across the Andes and pampas of Buenos Ayres, with an account of the minerals and mineral waters of Chili, their Indian antiquities, and a description of various mammals, birds, reptiles, fishes, crustacea, and plants collected by the expedition. Vol. III, will contain the differential observations made for the parallaxes of Mars and Venus at Santiago, Washington, Cambridge (Mass.), and Cape of Good Hope, together with a discussion of the question of parallax by Dr. Gould.   Vol. IV will contain the meridian circle observations of planets, time and azimuth stars, stars of Lacaille never re-observed by others, and a portion of the zone observations.   Vol. V. will contain the remainder of the zone observations.   Vol. VI. will embrace the magnetical and meteorological observations; and Vol. VII. will contain a catalogue of the fixed stars observed by the expedition, embracing determinations of position of more than 20,000 stars never before tabulated.

SECTION III.

VERY extensive surveys have been undertaken, at the expense of the general government, and some by State governments, which have indirectly contributed very much to the science of astronomy. Of these, the survey of the coast of the United States is the most important.

The survey of the coast was proposed by Mr. Jefferson, and was authorized by Congress in 1807. Mr. Gallatin, then Secretary of the Treasury, sketched the plan of a magnificent geodetic work in which the principal headlands of the coast should be fixed by astronomical observations. In consequence of the unsettled state of the country, no active steps were taken toward carrying this plan into execution until 1811, when Mr. Hassler was placed in charge of this work, and was sent to Europe to procure the requisite instruments. He did not return with the instruments until the fall of 1815. In 1816 he commenced the survey; and in 1818, Congress not being satisfied with the progress of the work, it was stopped.

In 1832, the work was revived by an Act of Congress, and placed under the direction of Mr. Hassler, in whose hands it made steady progress until his death in 1844. Professor A. D. Bache was then appointed to take charge of the survey, and has continued it to the present time.

The astronomical part of this survey consists in determining the latitude and longitude of the stations, and the direction of the sides of the triangles with reference to a meridian.

Professor Bache has undertaken to determine the difference of longitude between Greenwich and the most important points upon our coast, with the greatest possible precision. For this purpose he has availed himself of all the astronomical observations previously on record, and has instituted new observations, including those of occultations and eclipses, moon culminations, and the exchange of chronometers. Numerous observations have been made in Cambridge and its vicinity, in the neighborhood of New York, Philadelphia, Washington, and other places. The difference of longitude between these several places has been determined by means of the electric telegraph, so that observations at any of these places are equally available for determining the longitude of each of them from Greenwich.

Advantage has been taken of the frequent passage of steamers between Boston and Liverpool, to make a thorough comparison of the times of those ports by means of chronometers. For this purpose, as soon as a steamer arrives in Boston, its chronometers are taken to Cambridge observatory for comparison, where they remain until the steamer is ready to return. Upon arriving in Liverpool, the chronometers are taken to the Liverpool observatory, and their errors determined. This method of comparison has been systematically pursued since 1844. During the year 1846, forty-two such comparisons were made. In 1848 the longitude of the Cambridge observatory from Greenwich was determined by Mr. Bond, from the transportation of 116 chronometers, in thirty-four voyages of the Cunard steamers from Liverpool to Boston, to be 4h. 44m. 30·5s. The longitude deduced from lunar occultations and solar eclipses is 4h. 44m. 31·9. During the year 1849, eighty-seven additional comparisons were made, the results of which differ nearly two seconds of time from those previously obtained by astronomical observations. The mean result of 175 chronometers was 4h. 44m. 30·1s., and it was believed that this result could not be one second in error. The final result of the chronometric expeditions of 1849, 1850, and 1851 was 4h. 44m. 30·66s. During the progress of these expeditions, more than four hundred exchanges of chronometers have been made.

The results of the coast survey must furnish additional materials for determining the figure and dimensions of the earth. The Atlantic coast embraces more than twenty degrees of latitude, and its survey will virtually furnish a measured arc of the meridian of that extent. In several places the survey will furnish long continuous arcs upon the same meridian. Thus from Nantucket northward we shall obtain an arc of over three degrees; from Cape Lookout, northward along the shore of Chesapeake Bay, we shall obtain an arc of over five degrees; and the Florida coast will furnish us a continuous arc of more than seven degrees.

The survey of the boundary between the United States and Texas, in the year 1840, and the survey of the north-eastern boundary of Maine between the years 1840 and 1844, furnished the occasion for the determination of the latitude and longitude of numerous points, chiefly by Major J. D. Graham of the corps of Topographical Engineers.

In the summer of 1835, Captain Talcott was employed by the government of the United States, to make a series of observations near the southern line of Michigan, to settle the disputed question of boundary between that Territory and Ohio. In this expedition the latitude and longitude of several places were determined with great precision.

The most important astronomical survey hitherto undertaken by any State government was that commenced by the State of Massachusetts, in 1830, and completed in 1838. This survey was founded upon a base of 7·39 miles in length, measured on the banks of the Connecticut river, from which a net-work of triangles commenced and spread over the entire State. The latitude and longitude of twenty-seven places were determined independently by Mr. R. T. Paine; and the results of the two surveys agree remarkably with each other.

A topographical survey of the State of Maryland has recently been executed, under the direction of Mr. J. H. Alexander.

SECTION IV.

APPLICATION OF THE ELECTRIC TELEGRAPH TO ASTRO-NOMICAL USES.

In 1839, Professor Morse suggested to M. Arago, that the electric telegraph would afford the means of determining the difference of longitude between distant places with an accuracy hitherto unattainable.

The first practical application of this method was made by Captain Charles Wilkes, in June, 1844, between Washington and Baltimore.* Captain Wilkes conducted the experiments at Washington, and Lieutenant Eld at Baltimore. Two chronometers, previously rated by astronomical observations in the vicinity, were brought to the two telegraph offices, and were compared together through the medium of the ear, without coincidence of beats. The comparisons of the chronometers were continued for three days, and the results indicated that the Battle Monument at Baltimore was 1m. 34·87s. east of the Capitol. This method will furnish differences of longitude with a precision greater than any method hitherto practiced; but it is susceptible of great improvements.

---

* Vail's American Electro-Magnetic Telegraph, page 60.

In the year 1845, a plan was adopted by Professor A. D. Bache, Superintendent of the Coast Survey, to apply this method in an improved form, to the determination of the difference of longitude between the principal stations of the survey; and in 1846 measures were taken to connect in this manner Washington, Philadelphia, and New York. An arrangement was made with the telegraph company, to allow the use of their line for scientific purposes, after the usual business operations had closed for the day. A line of wires was extended from the General Post Office in Washington to the Naval observatory; a wire was carried from the main line through the High School observatory at Philadelphia; and a short wire was carried from the office in Jersey City to a station fitted up as a temporary observatory, and furnished with a five-feet transit telescope and an astronomical clock. The observations at Washington were made by Professor R. Keith; those at Philadelphia by Professor E. O. Kendall; and those at Jersey City by Professor E. Loomis; the whole being under the direction of Mr. S. C. Walker. Each station was furnished with a telegraph key and a receiving magnet.

Washington, Philadelphia, and Jersey City were thus put in telegraphic connection; instruments for obtaining time were provided; and to determine the difference of longitude of the stations, required simply the means of producing an instantaneous effect observable at all the stations. This was to be obtained by the motion of the armatures of electro-magnets, which had been previously adjusted by Mr. Saxton, so as to secure their simultaneous striking on the transmission of the electric current. The first trials which were made for the transmission of signals were unsuccessful. The observers were not provided with the means of holding communication by the ordinary mode of telegraphing, and if every thing was not arranged exactly as had been previously agreed upon, it was impossible to correspond for the purpose of discovering the source of the difficulty. Communication between Philadelphia and Washington was however effected on the 10th and 22d of October, and the difference of longitude approximately obtained. Signals for time by the clock were transmitted, and star signals were exchanged. On the 10th of October, the transit of the star 2838 Bailey over the seven wires of the west transit instrument of the Washington observatory was signalized by Lieutenant Almy. The time was noted on the Washington clock by Lieutenant Almy, and also by Mr. Walker, comparing together, by the ear, the seven key beats with the clock beats. The same key beats were also noted by Professor Kendall at Philadelphia. These observations gave for the difference of longitude between the two places 7 minutes and 34 seconds in time.

The experience of a few nights showed the necessity of complete registering apparatus, such as is ordinarily employed by the telegraph companies; and this was accordingly ordered, but was not received in season for use during the year 1846.

In the summer of 1847, the experiments were renewed with more complete apparatus. After 10 o'clock, P.M., the three stations at Washington, Philadelphia, and Jersey City, opposite New York, were converted into temporary telegraph offices, as well as astronomical stations. The astronomical observations for clock corrections and personal equations, were arranged by Mr. S. C. Walker, on the model of Struve's celebrated chronometric expedition between Pulkova and Altona.

The following is the method of comparing the local times at the different stations. A signal is given at the first station by pressing a key, as in the usual mode of telegraphing; and the observer at each of the other stations hears the click caused by the motion of the armature of his electro-magnet. The most obvious method of comparison consists in simply striking on the signal key at intervals of ten seconds; the party at the first station recording the time when the signals were given, and the party at the second station recording the time when the signals were received. After about twenty signals have been transmitted from the first station to the second, a similar set of signals is returned from the second station to the first. The objection to this mode of comparison is that it requires the fraction of a second to be estimated by the ear. The party giving the signals strikes his key in coincidence with the beats of his clock, so that at his station there is no fraction of a second to be estimated; but at the other station, the armature click will not probably be heard in coincidence with the beats of the clock, and the fraction of a second is to be estimated by the ear. Now this fraction can not be estimated with the accuracy which is demanded in this kind of comparison. It is found that observers generally estimate the fraction of a second too small when using the ear alone, unassisted by the eye. This error is greatest at the middle date between two clock beats, and is found to vary from 0·06 to 0·18 of a second with different observers.

The experience of a few nights with the preceding method of observation, suggested a second method of comparison which relies on the coincidence of a mean solar and sidereal clock or chronometer. A sidereal clock gains upon a solar clock one second in about six minutes; and if two such clocks are placed side by side they must tick together once in every six minutes. In order to compare two such clocks, we notice their movements, and wait until the beats sensibly coincide, when we know that their difference amounts to an entire number of seconds, which is readily discovered. Chronometers generally make two beats in a second; so that between a clock which beats seconds of sidereal time, and a chronometer which ticks half seconds of solar time, there must be a coincidence every three minutes.

The following is the method pursued in the comparisons between Philadelphia and Jersey City. After transmitting a few signals by the former method, so as to determine the difference between the local times of the two stations within a small fraction of a second, the party at the first station commences striking on his signal key every second, in coincidence with the beats of his mean solar chronometer, and continues to do so for ten or fifteen minutes without interruption. The party at the second station compares the armature click of his magnet with the beats of his sidereal clock, and watches for a coincidence, and records the time when a coincidence takes place. When he has obtained two or three coincidences, which generally requires from ten to fifteen minutes, he breaks the electric circuit, in order to notify the first party to stop beating. He then commences beating seconds by striking his own signal key in coincidence with the beats of his sidereal clock; and the party at the first station compares the armature clicks of his magnet with the beats of his solar chronometer, and watches for a coincidence. When he has obtained three or four coincidences, which generally requires ten or twelve minutes, he breaks the electric circuit, in order to notify the other party to stop beating. The comparison of times at the two stations is now complete.

The following is the result of this summer's campaign :

	DIFFERENCE OF LONGITUDE BETWEEN		
	Philadelphia and Washington.	Philadelphia and Jersey City.	Washington and Jersey City.
1847, July 19,	7m. 32s.969	4m. 30s.440	
" 21,	33.195		
" 24,	33.058	30.305	
" 27,		30.425	
" 28,		30.470	
" 29,	33.050	30.407	12m. 3s.552
Aug. 3,	33.134	30.386	3.452
" 10,		30.439	
" 11,		30.303	
Means.	7m. 33s.079	4m. 30s.397	12m. 3s.504

For the distance of 250 miles embraced in these experiments, the electric current took no sensible time to
propagate itself, and it appeared that two clocks at this
distance could be compared with the same degree of
precision as if they were placed side by side.

During the summer of 1848, similar experiments were
made to determine the difference of longitude between
New York and Cambridge. A wire was extended from
the Cambridge observatory, to connect with the New
York and Boston line near Brighton; and another wire
was carried from the same line to Mr. Rutherford's observatory in the upper part of New York city. Thus
the observatories at New York and Cambridge were
put in telegraphic communication. At New York a
new forty-five inch transit instrument, by Simms, of
London, belonging to the Coast Survey, was used for
local time, and a sidereal clock with chronometers for
comparison. At Cambridge a similar transit was used,
with numerous chronometers carefully compared. The
observations at Cambridge were made by Professor
W. C. Bond, and those at New York by Professor
E. Loomis. The comparisons of time were made both
by the method of coincidences already explained, and
by telegraphing the transit of the same star over both
meridians. The latter method was practiced in the following manner:

A list of zenith stars is selected beforehand, and
furnished to each observer. When every thing is prepared for observation, the Cambridge astronomer points

his telescope upon one of the selected stars as it is passing his meridian, and strikes the key of his register at the instant the star appears to coincide with the first wire of his transit. He makes a record of the time by his own chronometer; and the New York astronomer, hearing the click of his magnet, records the time by his own clock. As the star passes over the second wire of the transit instrument, the Cambridge astronomer again strikes the key of his register, and the time is recorded both at Cambridge and New York. The same operation is repeated for each of the other wires. The Cambridge astronomer now points his telescope upon the next star of the list, which culminates after an interval of five or six minutes, and telegraphs its transit in the same manner. In about twelve minutes from the former observation, the first star passes the meridian of New York, when the New York astronomer points his transit instrument upon the same star, and strikes the key of his register at the instant the star passes each wire of his transit. The times are recorded both at New York and Cambridge. The second star is telegraphed in a similar manner. The Cambridge astronomer now selects a second pair of stars, and repeats the same series of operations, and is followed by the astronomer at New York, when the star comes upon his own meridian. By this comparison, the difference of time between the two stations is obtained independently of the tabular places of the stars.

On seven nights in July and August these methods

were practiced, during which time 12,000 signals were
exchanged between the observatories at New York and
Cambridge.

The personal equations for the clock corrections of the
different observers were obtained by a very extensive
series of comparisons, 894 in number, the transit of the
same star over alternate wires of the telescope being
noted by different observers.

The result of all the comparisons gave the difference
of longitude between Cambridge and Mr. Rutherford's
observatory, 11 minutes and 26.07 seconds.

During the month of October, 1848, the difference of
longitude between the Cincinnati observatory and the
High School observatory in Philadelphia was also determined by telegraph for the use of the United States
Coast Survey. A wire was carried from the Cincinnati
observatory to the Philadelphia line, thus putting the
observatories of Philadelphia and Cincinnati in telegraphic communication. The series of observations here
made was substantially the same as had been practiced
two months before between New York and Cambridge.
A sidereal clock in Philadelphia was compared with a
solar chronometer in Cincinnati, by beating seconds upon
the key of the telegraph register for fifteen minutes, and
noting the instants of coincident beats. Transits of the
same stars over both meridians were also telegraphed,
as had been practiced between Cambridge and New
York. These comparisons were made on six different nights, and gave the difference of longitude be-

tween Philadelphia and Cincinnati 37 minutes 20.48
seconds.

In the course of the comparisons for longitude by telegraph, up to the autumn of 1848, many thousand signals
were transmitted, and all by the hands of a human
operator. But it is impossible for human fingers to
move with the precision of machinery; and, after the
first successful trial of the telegraph for longitude, it
became evident that an important advantage would be
secured if the clock could be made to transmit its own
signals. The desideratum was to make an astronomical
clock break the electric circuit every second, so that practically it might be said that its beats could be heard along
the entire line of telegraph communication; and this
must be done in such a manner as not to affect the rate
of the clock. A number of different methods of ac-
complishing this object were speedily proposed; but as
several contrivances of a similar kind had previously
existed, it is thought best to notice them all in chrono-
logical order.

THE ELECTRIC CIRCUIT BROKEN BY A CLOCK.

The first invention for breaking the electric current by
clock-work appears to be due to Professor Steinheil, of
Munich, who previous to September, 1839, had perfected
a method for causing any number of clocks to indicate
exactly the same time.* This was accomplished by
means of an arrangement which enabled the regulating

---

* Moigno's Traité de Telegraphie Electrique, page 337.

clock, at the end of every hour, to advance or put back the hands of the subordinate clocks, so that all should indicate exactly the same time. This contrivance was as follows: One of the wheels in each clock carries a flat piece in the form of a spiral, which during the hour, slowly raises a weight acting on a lever. The weight, when raised, is sustained like the trigger of a musket, and the spiral piece has a notch in which the lever may be caught. When the instant arrives for the standard clock to regulate all the others, an electro-magnet (rendered magnetic by the instantaneous passage of an electric current) attracts its armature, causing the arm of the lever to fall, and in its fall it catches in the notch of the spiral piece, with which the hands of the clock are connected. If during the preceding hour, the clock has gained or lost time, the fall of the lever carries backward or forward the spiral piece, and with it the hands of the clock, so that on every dial the hands indicate exactly the same time. Thus all the clocks of a large city may be made to strike the hour at the same instant.

For the ordinary purposes of business, it is not necessary that the agreement of the clocks should extend to very minute fractions of time; if this was however desired, the standard clock might act directly upon the pendulums of the subordinate clocks, so as to render their times of vibration perfectly equal.

In the year 1840, Professor Wheatstone, of London, invented an apparatus, called an electro-magnetic clock, for enabling a single clock to indicate exactly the same time in as many different places as may be required.* In the electro-magnetic clock, all the parts employed in an ordinary clock for maintaining and regulating the power are entirely dispensed with. It consists simply of a face with its second, minute, and hour hands, and of a train of wheels which communicate motion from the arbor of the second's hand to that of the hour hand, in the same manner as in an ordinary clock train. A small electro-magnet is caused to act upon a peculiarly constructed wheel, placed on the second's arbor, in such a manner that whenever the temporary magnetism is either produced or destroyed, the wheel, and consequently the second's hand, advances a sixtieth part of its revolution. It is obvious then that if an electric current can be alternately established and arrested, each resumption and cessation lasting for a second, the instrument now described, although unprovided with any internal maintaining or regulating power, would perform all the usual functions of a perfect clock. The manner in which this apparatus is applied to the clocks, so that the movements of the hands of both may be perfectly simultaneous is the following:

On the axis which carries the scapement wheel of the primary clock, is fixed a small disc of brass, which is first divided on its circumference into sixty equal parts; each alternate division is then cut out and filled with a piece of wood, so that the circumference consists of thirty regular alternations of wood and metal. An ex-

---

* Abstracts of the Philosophical Transactions vol iv., page 249.

tremely light brass spring, which is secured to a block of
ivory or hard wood, and which has no connection with
the metallic parts of the clock, rests by its free end on
the circumference of the disc. A copper wire is fastened
to the fixed end of the spring, and proceeds to one end
of the wire of the electro-magnet; while another wire
attached to the clock-frame is continued until it joins the
other end of that of the same electro-magnet. A constant voltaic battery, consisting of a few elements of very
small dimensions, is interposed in any part of the circuit.
By this arrangement, the circuit is periodically made and
broken, in consequence of the spring resting for one
second on a metal division, and the next second on a
wooden division. The circuit may be extended to any
length; and any number of electro-magnetic instruments
may be thus brought into sympathetic action with the
standard clock.

In the year 1840, Mr. Alexander Bain, of London, invented an arrangement by which it was proposed to work
a great number of clocks simultaneously.* The following is the method by which the regulating clock was
made to break the electric circuit. A B is a pendulum
vibrating seconds. C is a plate of ivory affixed to the
frame of the clock, in the middle of which is inserted
a slip of brass D, communicating with the positive pole
of a voltaic battery. To the pendulum is attached a very
light brass spring E in such a manner that every vibra-

---

* Applications of the Electric Fluid to the Useful Arts, by Mr.
Alexander Bain, page 97.

tion of the pendulum brings the free end of the spring
into contact with the strip of brass D, thus completing
the electric circuit through the
upper part of the pendulum to the
negative pole of the battery; while
the circuit is broken as soon as the
spring touches the ivory.

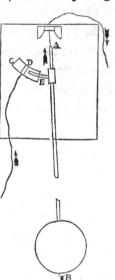

Subsequently Mr. Bain adopted
the arrangement of closing and
breaking the circuit by means of a
short brass bar, placed in a horizontal position near the middle of
the pendulum; the bar being slid
back and forth about one inch at
each vibration of the pendulum.

In 1844, Mr. Joseph Saxton
proposed to break the electric
circuit by a pin on the pendulum rod striking a small
tilt hammer, and also making the circuit by a lancet-
shaped point of platinum at the bottom of the rod
passing through a globule of mercury. Early in 1846,
he proposed both methods again, when objections were
made against them on the supposition that the current of
electricity passing through the pendulum rod might affect
the rate of the clock. To meet this objection, he proposed to insulate the pin by a piece of ivory, or use a
glass pin. Or if the other method was used, instead of
passing the current through the rod, he proposed to at-
tach at the top of the rod a small lever to be raised by

the motion of the pendulum; the lever to turn on a
center near the point of suspension, but insulated
from it.

In 1849, the arrangement with a glass pin moving a
platinum tilt-hammer was applied by Mr. Saxton to the
Hardy clock of the Coast Survey, and put in operation at
the Seaton station in July of that year. This arrangement is shown in the annexed cut. A B C represents a

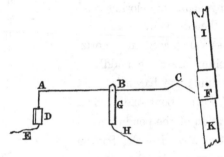

fine platinum wire, mounted on a pivot at B, the end A
being somewhat heavier than the other, and resting upon
a metalic bed D. At C the wire is bent so as to form an
obtuse angle. The wire E goes from D to one pole of
the battery, while the wire H from the other pole of the
battery, communicates with the metallic support G, and
thence with the wire A B. When the end A of the
platinum wire rests upon the support D, it is evident that
the electric circuit is complete. This apparatus is placed
near the middle of the pendulum (a portion of which,
I K, is represented in the cut), and just in front of it, so
that the pendulum may swing behind it without obstruction. A small glass pin F, about half an inch in length,

is attached to the pendulum in such a position that, at
every vibration of the pendulum, the pin slightly im-
pinges upon the angle C, of the platinum wire and forces
up the end A. As soon as the pin has passed the point
C, the end A falls back again upon its support D.   Thus
at every vibration of the pendulum, the end of the platinum wire is lifted about a tenth of a second, and rests
upon D during the remaining nine tenths of the second;
that is, the electric circuit is closed about nine tenths of
every second, and is open during the remaining tenth.

The method first proposed by Mr. Saxton in 1844, is
the one which has been employed at the Washington
observatory since 1849.   A small piece of metal M is
attached to the back of the clock, near
the lower extremity of the pendulum,
and upon it is placed a small globule of
mercury, so that the index B attached to
the lower extremity of the pendulum may
pass through the globule of mercury once
in every vibration.   A wire from one
pole of the battery is connected with the
supports of the pendulum C, and a second
wire from the other pole of the battery
connects with the metallic support of the
mercury globule.   If now the pendulum
were at rest, with the point B in the
mercury, it is evident that the electric
circuit would be complete through the
pendulum.   If then the pendulum be set in motion, it

will break the circuit whenever it passes out of the mercury, and restore it again as soon as it touches the mercury.

In 1843, at the launch of the frigate Raritan at Philadelphia, an attempt was made to ascertain the rate of motion of the ship when running off the ways, by observing the time required for certain marks on the side of the ship to pass fixed points of sight. As this method proved unsatisfactory, Mr. Saxton contrived a machine for registering the motion with certainty. This contrivance consisted of a half second's pendulum, with a pin projecting from the rod. This pin acted on an angle piece attached to a lever which had a small conical cup at the end, containing a mixture of oil and vermilion. The cup had a small hole at its apex, and the mixture was prevented from flowing out by the capillary action of a small lock of cotton. A reel, containing about 60 feet of white cotton tape, was so placed that the tape passed under the cup as it was drawn out, and had a red dot struck on it every half second by the vibration of the pendulum.

Some time afterward, when the steam frigate Princeton was ready to be launched, Mr. Saxton constructed a machine of the kind here described. A wire about two feet long was attached to a small ring that hooked on a detent, which held the pendulum when it was raised up on one side. The other end of the wire was attached to the bow of the ship, so that the first motion of the ship would unhook the pendulum, which, falling to its lowest point, caused the lever to strike a dot on the tape at the first quarter second, and afterwards at each half second to the end of the tape. This pendulum had no maintaining power, except its own gravity, as it was only required to act for a few seconds. Electricity was not used in this experiment; but this machine was the germ of the contrivances described on page 318 for breaking the electric circuit.

In the year 1847, Mr. J. J. Speed, of Detroit, Michigan, conceived a plan for causing all the clocks of a large city to indicate the same time. He proposed to have all the clocks in the city connected with galvanic circuits, and operated from some central battery, the hands on the clocks moving only at given intervals, and at the instant the circuit should be closed. To close and break the circuit, he had a clock constructed with a tilt-hammer, which was lifted by a projecting tooth on one of the wheels of the clock. The credit of this part of the invention is conceded, however, to Mr. C. F. Johnson, of Owego, N. Y., and was patented by him in 1846. The idea of applying the galvanic circuit to give motion to house clocks by means of the tilt hammer, is claimed by Mr. Speed. His attention being soon afterward directed to other objects, Mr. Speed never carried his plan into execution, and the clock which he ordered to be constructed with the tilt-hammer arrangement is now the property of the United States Coast Survey.

Dr. Locke, of Cincinnati, in the autumn of 1848, invented an arrangement for breaking the electric circuit

14*

by means of a tilt-hammer.   Dr. Locke employs a wheel
with sixty teeth attached to the axis of the escapement
wheel.   Each tooth in succession strikes against the
handle of a platinum tilt-hammer, A C, weighing about
two grains, and knocks up the hammer, which almost
immediately falls to a state of rest on a bed of platinum.

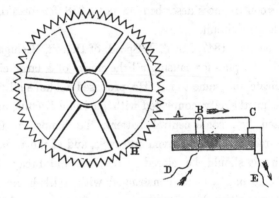

The fulcrum B of the tilt-hammer and·the platinum
bed rest severally on a small block of wood.   Each is
connected by wires D and E with a pole of the galvanic
battery, and the circuit is alternately broken and completed by the rising and falling of the hammer.   The
circuit is open about the one tenth of a second, and
closed the remaining nine tenths of each second.   This
arrangement was first tested on the 17th of November,
1848, on the Cincinnati and Pittsburg line, about four
hundred miles in length.   The circuit was broken every
second by the motion of the clock; and the fillet of
paper being allowed to run off from the reel of the telegraph register, it was graduated into equal portions,

consisting of an indented line about nine tenths of an inch in length, followed by a blank space of about one tenth of an inch.

The Congress of the United States have expressed their conviction of the importance of this invention, by awarding the sum of ten thousand dollars to Dr. Locke for his invention, and directing that a clock upon this principle should be constructed for the use of the observatory at Washington. This clock was completed in 1850, and has been set up for use at the observatory.

In 1849 Professor O. M. Mitchell, of Cincinnati, invented a method of breaking the electric circuit by means of a delicate fibre attached to the pendulum which acts upon a cruciform lever, and thus, in every vibration of the pendulum, allows a metallic point to dip into a cup of mercury, which completes the circuit.   In the annexed figure, A B represents the lower extremity of the pen-

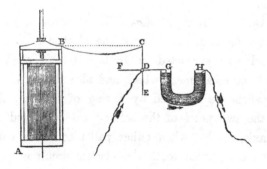

dulum ; B C is a delicate fibre, one end of which is attached to the pendulum, and the other to one arm of a cross, C E F G, formed of platinum wire, mounted like a wheel upon the axis D. The arm D G of the cross being slightly the heaviest, rests upon one end of a glass tube, G H, bent in the form of a syphon, and containing mercury, while the other end, H, of the tube is without the clock case, so that it can be reached without opening the case. The axis D of the cross communicates by a wire with one pole of the battery, while the end H of the mercury tube communicates with the other pole. As the pendulum approaches to the extreme left of its arc of vibration, it pulls, by means of the fibre B C, upon the arm C D, and lifts the end G out of the mercury, thus breaking the electric circuit; but during the remaining part of each double vibration, the point G rests upon the mercury, and the electric circuit is complete. Accordingly, in Professor Mitchell's register, the clock dots are made at intervals of two seconds.

In October, 1848, Professor W. C. Bond made the drawings for a clock to break the electric circuit, and his clock was completed in 1850. In this clock the axis of the escapement wheel, and also the axis of the steel pallets are insulated by a ring of shellac. Wires from the two poles of the battery are connected with each axis, so that when either pallet comes in contact with an escapement tooth, the electric circuit is closed; and when the contact is broken (as it must be at every oscillation of the pendulum) the electric circuit is opened.

MODE OF REGISTERING THE OBSERVATIONS.

The most obvious mode of registering the beats of the clock is upon a long fillet of paper, after the ordinary method of telegraphic communications. If the paper be allowed to run through an ordinary Morse registering apparatus, and the circuit be broken every second by the clock, the graver will trace upon the paper a series of lines of equal length separated by short interruptions thus :

It is easy to reverse the action of the graver, so that when the circuit is complete, the paper shall be entirely free, and a dot be made by the breaking of the circuit. A paper graduated into seconds by this arrangement exhibits dots with long intervening spaces thus :

instead of long lines with short blanks, as shown before.

In order to indicate the commencement of the minute, a dot may be omitted at the end of every 60 seconds. This is accomplished in Dr. Locke's clock by omitting one tooth in the wheel which breaks the circuit, as shown at H in the figure page 322.

The mode of using the register for marking the date of any event, is to tap on a break-circuit key, simulta-

neously with the event. The beginning of the short line
thus printed upon the graduated scale of the register,
fixes, by a permanent record, the date of the event.
Thus A represents such a record printed upon the graduated paper :

—   —   —A—   —   —   —

By tapping upon the key at the instant a star is seen
to pass each of the wires of a transit instrument, the observation is instantly and permanently recorded.   The
usual rate of progress of the fillet under the pen is about
one inch per second, and the observations are read off by
means of a graduated transparent scale, about an inch
square, as represented in the annexed cut, consisting of

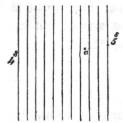

equidistant and parallel lines, ruled
upon a piece of glass, by means of
a diamond or etched with fluoric
acid. If the interval between the
second dots be greater than the
breadth of the scale, the scale is
turned obliquely across the fillet,
until the first and last divisions exactly comprehend the
space between the two second dots.   Let the distance
from 4s. to 5s. on the above scale, be the distance on the
fillet between the fourth and fifth seconds, and let the dot
a between them represent the observation. It appears,
by inspection, that the observation was recorded between
4·7 and 4·8 seconds. The distance of a from the nearest
scale division may be estimated to tenths. Thus time is

accurately measured to tenths, and may be estimated to hundredths of a second.   On some ac-counts it is more convenient to employ a scale consisting of diverging lines, as represented in the annexed cut, so that the breadth of the scale may always exactly comprehend the interval be-tween the second dots, which intervals must necessarily vary somewhat in length.

It is important that the paper upon which the observa-tions are recorded should be reduced to a convenient form for preservation.   The long fillet of paper employed in ordinary telegraphing, is very inconvenient for astro-nomical purposes.   If the paper is allowed to run off at the rate of one inch per second, the length of fillet re-quired for one hour's observations would be 3600 inches, or 300 feet; and for a single night's work of an observatory, a length of nearly half a mile would be required.   The inconvenience of managing such a strip of paper detracts materially from the value of the method.

In the summer of 1849, Mr. Saxton completed a re-gister which is well adapted to the regular work of an observatory.   It consists of a cylinder which may be made of any convenient dimensions, say six inches in diameter and two feet long, enveloped with paper which may be removed at pleasure.   This cylinder is made to revolve with a uniform motion, while the registering pen moves forward at the rate of one tenth of an inch to every revolution of the cylinder.   Thus the pen is made to trace a spiral on the cylinder, and one sheet of paper, twelve inches by twenty, lasts for more than two hours of constant work.   Two sheets will contain an ordinary night's work.

In order to secure the full advantage of this method, it is important that the paper which contains the register be made to advance with entire uniformity.   The Messrs. Bond have invented for this purpose a machine which they call the Spring Governor, consisting of a train of clock-work connected with the axis of a fly-wheel.   It has an escapement-wheel, into the teeth of which pallets are operated by the oscillations of a pendulum, as in ordinary clocks, the wheel being so connected with its axis by a spring as to allow the axis to move while the wheel is detained by the pallets.   The register is made upon a sheet of paper wrapped round a cylinder.   The annexed specimen is a fac-simile, taken from a record sheet used at the Cambridge observatory. A B shows the signal announcing the approach of a star to the right ascension wires of the transit instrument; C indicates the passage of the star over the first wire; and D indicates the passage over the second wire

The passage over the first wire took place at 7h. 16m. 7s.4, and the passage over the second wire at 7h. 16m. 11s. 4.

Mr. Kerrison has invented a method of regulating the motion of the registering cylinders by means of two short pendulums attached to cranks, in such a manner, that the one crank shall be at its maximum effect when the other is passing its dead point. The weight, by which the works are kept in motion, has in this arrangement a considerable influence on the velocity; consequently it was found very difficult to regulate it, and when regulated, almost impossible to keep it so. The number of pendulums were increased by Mr. Kerrison in the hope of overcoming the difficulty, but hitherto without success.

Professor Mitchell's method of recording right ascensions was invented in 1849, and consists of a horizontal disc which is made to revolve with uniform velocity once a minute on a vertical axis. This disc carries either a metal plate, or a paper disc, on which the time and observations are recorded by pens drawn by electro-magnets. At every alternate second, a dot is struck by the time-pen on the paper. When the disc has performed one revolution, a tooth upon the axis of the disc takes hold of a fixed rack, and moves the traveling frame, which carries the center of the disc through the tenth of an inch, when a new circumference of time dots is. commenced. The observation dots fall intermediate between the minute circles and second dots, and are struck so as to distinguish them by form from the time dots.

OBSERVATIONS FOR LONGITUDE SINCE THE AUTUMN
OF 1848.

On the 17th of November, 1848, Professor Locke, at
Cincinnati, undertook so to connect his clock with the
telegraph line that its beats should be heard and registered at Pittsburg, a distance of about four hundred
miles. The circuit was broken every second at Cincinnati by the motion of the clock; and at Pittsburg,
the fillet of paper being allowed to run off from the reel
of the telegraph register, it was graduated into equal
portions, consisting of an indented line about nine-tenths
of an inch in length, followed by a blank space of about
one-tenth of an inch. The two correspond to one second
of time, commencing with the beginning of the line.
This experiment was continued for two hours, during
which time the seconds of the Cincinnati clock were
registered on the running fillet of paper at all the offices
along the line. In order to distinguish the hours and
minutes upon this graduated paper, Dr. Locke proposed
to make the beginning of the ordinary minutes omit
one blank space; the beginning of five minutes omit *two*,
of ten minutes *three*, and of an hour omit *four* consecutive blank spaces. Thus, ordinary beginnings of
minutes have continuous lines of *two* seconds, fives *three*,
tens *four*, and hours *five* seconds.

In the month of January, 1849, a clock upon Dr.
Locke's construction was employed for printing transits
of stars over different meridians for the determination

of longitude.   The observatories at Cambridge, New
York and Philadelphia, were all put in communication
with each other, and with Washington city.   The clock
which was to be employed was set up in Philadelphia,
and connected with the telegraph line.   Simultaneously
with the beats of this clock, a click was heard of the
magnets at Cambridge, New York and Washington.
The paper being allowed to run off from the reel, it was
graduated into parts corresponding to the beats of the
Philadelphia clock.   The astronomer at Cambridge now
selects a convenient star for observation, and announces
it by name to each of the other stations.   He strikes the
key of his register as the star passes successively each
wire of his transit instrument, and the dates are printed
not only upon his own roll of paper, but also upon those
at New York, Philadelphia and Washington.   When
the same star comes over the meridian of New York,
the observer there goes through the same operation, and
his observations are printed upon all four of the papers.
The Philadelphia observer does the same when the star
comes upon his own meridian.   Thus we have four long
rolls of paper, one at Cambridge, a second at New York,
a third at Philadelphia, and a fourth at Washington, all
graduated into equal parts by the ticking of the Philadelphia clock, and upon these we have printed the
instants at which the star was seen to pass each wire of
the transits at Cambridge, New York and Philadelphia.
The position of each mark thus printed shows not only
the second of occurrence, but also the fraction of a

second, which may be measured with scale and dividers.
Thus, if we suppose the transit instruments to be all adjusted to the meridian, and the rate of the clock to be
correct, we have obtained the difference of longitude of
the stations compared, independently of the tabular place
of the star employed, and also independently of the
absolute error of the clock. The observers now read
their levels, and reverse their transit instruments. The
Cambridge astronomer selects a second star, which is
telegraphed in the same manner as the first. Thus the
error of collimation of all the telescopes is corrected.
The other errors of the instruments must be determined
by separate observations in the usual manner. Subsequently a third and a fourth star were selected by the
Cambridge astronomer, and telegraphed in the same
manner as they passed in succession over the different
meridians. These experiments were made on the 23d
of January, 1849, and occupied most of the night. The
results were most wonderful, and have opened an entirely new field of investigation. Hitherto in transit
observations, astronomers had been accustomed to estimate fractions of a second entirely by the ear, with
only such assistance from the eye as could be derived
from the rapid motion of the star through the field of the
telescope. The error of such an observation, even with
practiced observers, frequently amounts to a quarter of
a second. But in this new mode of observation, the
observer has no use for his ears. The astronomer
might in future be made without ears. It is only nec-

essary for him to move his fingers at the instant the star is seen to pass each wire of his telescope, and his observation is recorded in a permanent form, and may be subsequently examined at his leisure.

During the months of July and August, 1849, a telegraphic comparison was made between the observatories of Philadelphia, and Hudson, Ohio. Signals were exchanged on three different nights, the results of which gave the difference of longitude between the High School observatory, and the Hudson observatory 25m. 5·7s. During the same summer, some new comparisons were made between Philadelphia and Washington.

On the 5th of February, 1850, a perfect telegraph connection was made between Washington and Charleston, South Carolina. The thermometer at Washington was 10° Fahrenheit, and at Charleston 28° Fah., and the insulation of the wires was excellent. Professor Lewis R. Gibbes was in charge of the operations at Charleston, and the operations at Washington were conducted by S. C. Walker. The times of transit of a full series of zenith stars were telegraphed between Professor Gibbes's observatory and Washington, and recorded on the register at each station. The instruments were all well adjusted, and all necessary precautions taken to insure satisfactory results. Signals were also exchanged on the 11th and 12th of February, but the results were not so satisfactory as those of the 5th. The result of these comparisons indicates that Charleston is 11m. 45·27s. west of Washington.

In March, 1851, the difference of longitude between Charleston and Savannah was determined by telegraph. At the Charleston end, the observations were made by Professor Lewis R. Gibbes, at his observatory; and at the Savannah end by Mr. C. O. Boutelle. Thus the Seaton station at Washington was connected with Savannah, the difference of longitude between Charleston and Washington having been determined in the operations of the previous year.

In December, 1851, was determined the difference of longitude between Cambridge, Mass., and Bangor, Maine; and also between Bangor and Halifax, Nova Scotia, under the general direction of Professor S. C. Walker. Professor Bond, of Cambridge, and Captain Shortland, of the British Admiralty Survey of Nova Scotia, had concerted the connection of these two places by the telegraph; and it was considered important to furnish an intermediate station in Maine, and to connect the Survey of the United States with the British Survey. The operations were entirely successful, as far as Cambridge and Bangor were concerned, but the number of signals exchanged between Bangor and Halifax was small.

The difference of longitude between Seaton station, at Washington, and Roslyn station, near Petersburg, Va., was determined by observations made on six nights between July 3 and August 7, 1852. The chief object in these observations was to examine the influence of the different circumstances producing errors in the longitude determinations by this method. In determining the difference of longitude, one hundred and twenty-four observations were made upon thirty-five zenith stars; eighteen observations for collimation and nine for equatorial intervals were made upon three circumpolar stars. In connection with these operations, one hundred and twenty-two observations were made upon twelve stars for local time. The instrument employed was a forty-three inch transit instrument; the diaphragm consisted of twenty-five wires arranged in groups of five. The result of these observations shows Petersburg to be 1m. 35·603s. west of Washington, with a probable error of ± 0·009s. The probable error of a single set of observations of a star over fifteen wires is ± 0·081s. The residual probable error of a single night's work, with the transits of fifteen stars telegraphed from and received at both stations, not accounted for by the error of tapping and receiving, is but 0·004s., or may be considered insensible, so that it is unnecessary to multiply the number of nights of observation.

During the winter of 1852 and '53, a second determination of the difference of longitude between Charleston, South Carolina, and the Seaton station in Washington was attempted under the direction of Dr. B. A. Gould. From the middle of December until the middle of February, Dr. Gould remained in Charleston, observing on every fair night circumpolar stars with reversals, and both zenith and equatorial stars chronographically, for the determination of time and instrumental corrections. Whenever the sky was unclouded both at Washington and Charleston, the batteries were set up, and the greater part of the night spent in attempts to obtain direct communication between the two places. After a series of the most varied experiments, the operators were forced to the conclusion that the condition of the telegraph wires was such that direct telegraphic communication between the two stations was impossible. It thus became necessary to establish an intermediate station: and one was accordingly erected at Raleigh, North Carolina. It was here found that direct communication both with Washington and Charleston was possible. The series of observations was completed on the 14th of May, 1853, at which time eighty-four transits of stars had been exchanged with Washington on four nights; and fifty-nine exchanged with Charleston on four nights.

During the winter of 1853 and '54, the longitude experiments were renewed under the direction of Dr. Gould. On account of the telegraph facilities, it appeared desirable that Columbia, South Carolina, rather than Charleston should form a link in the great chain of telegraphic longitude-stations which are to connect the north-eastern with the south-western sea-ports of the United States, and a station was accordingly selected at Columbia. The observations at Columbia were made by Dr. Gould and those at Raleigh by Mr. G. W. Dean. It was Dr. Gould's intention to push the longitude connections as far as Macon, Gorgia, and an astronomical station was also selected in that city, but the unfavorable state of the weather prevented his occupying that station.

The observations were commenced at Columbia, January 4, 1854, but no opportunity for exchanging signals with Raleigh could be obtained until January 21st. It was not until the 11th of February, that three good series of star signals had been exchanged. Dr. Gould then took charge of the station at Raleigh, while Mr. Dean went to Columbia, and on the 12th of March, they succeeded in completing the second series of three nights' satisfactory exchange of signals.

The season being too far advanced for the proposed connection of Columbia and Macon, the Columbia instruments were removed to Wilmington, North Carolina, and those at Raleigh to Roslyn station near Petersburg, Va., and the observations for the connection of these two places commenced with the month of May. Star signals were exchanged between Mr. Pourtales at Petersburg and Mr. Dean at Wilmington on the 27th of May and 6th of June, when they exchanged stations, and again worked successfully on the 20th and 23d of June.

In January and February, 1855, observations were made to determine the difference of longitude between Cambridge, Mass., and Fredericton, New Brunswick. It was originally intended to have an unbroken telegraph communication between the Fredericton observatory and that of Harvard University, but in consequence of the wires from the latter to the office in Boston being out of repair, Professor Bond found it necessary to trust to two sidereal chronometers for the interval. The chronometers were carefully and repeatedly compared with

15

the transit clock at Cambridge, both before and after in-
terchanging signals, so as to ascertain their error and
rate; and at both observatories, on each day of operations,
the meridian passages of a number of stars were observed,
in order to obtain the error and rate of the transit clocks.
The method of operation was as follows: The assistant
at Boston commenced at an even minute by his chronometer, and sent second-beats for fifty consecutive
seconds. This was continued for ten successive minutes,
beginning always at the even minute, and the times were
noted by the transit clock at Fredericton. Subsequently the observers at Fredericton took the initia-
tive and sent a series of signals to Boston. The following
are the results of these comparisons:

1855, Jan. 23.	Longitude by signals sent from B. to F.	17m. 57·38s.
Feb. 2.	"	F. to B.	57·30s.
"	"	F. to B.	57·30s.
"	"	B. to F.	57·20s.
Feb. 10.	"	F. to B.	57·14s.
"	"	B. to F.	57·19s.
"	"	F. to B.	57·19s.
"	"	B. to F.	57·19s.

Result of all the comparisons 17m. 57·23s.

During the winter of 1854–5, the telegraph operations
for longitude were extended as far as Macon, Ga., and
arrangements were made for continuing the work to
Montgomery and Mobile. In December, 1854, Dr.
Gould, assisted by Mr. Goodfellow, occupied the observatory at Columbia, S. C., and Mr. Dean took charge

of the station at Macon. After successful exchanges of signals on three different nights, Messrs. Dean and Goodfellow interchanged stations, and obtained exchanges on three more nights with satisfactory results. The series of astronomical observations extended from December 30th to March 16th. The clocks employed were adapted to the chronographic method of observation by Mr. Saxton. The spring governor and Kerrison regulator were used for the chronographic registry of the transit observations; but during the exchange of telegraphic signals, an ordinary Morse register was also employed at each station, and was found to give results comparable in accuracy with either of the former, owing to the greater length of the seconds as recorded upon the fillet, and to the general uniformity maintained during any change of rate.

During the winter of 1855-6, a comparison was made between Wilmington, N. C., and Columbia, S. C., by means of a new telegraph line, which gives one determination of longitude between Petersburg and Columbia by the way of Raleigh, and another by the way of Wilmington. Signals were also exchanged between Macon and Montgomery, Ala. Preliminary arrangements have been made at Mobile for the next season's work, and the station at New Orleans has been selected. The arrangements for determining the difference of longitude between New Orleans and the observatories along the eastern coast of the United States are therefore fast approaching their completion.

The following are the rules now adopted in the Coast Survey operations for longitude. After the operators have connected the observatories and adjusted their magnets, so that the two stations receive each other's writing well with the least possible pass, the clock of the most eastern station is put on. The observer at the other station then strikes a dot each alternate second, until the observer at the clock station signifies "Aye, aye," by double dots, each alternate second. The exchange of signals then begins as follows:

First star.

Reading of level.

Second star.

Reversal of instrument.

Third star.

Reading of level,

etc.   etc.

In general it is only desirable to observe on the three middle tallies. The observer always informs the recorder of the tallies observed, and of any lost or badly struck threads. When ten stars have been satisfactorily exchanged, the eastern clock is taken off the circuit, and the western clock put on; and the exchange of ten stars more completes the telegraphic work for the night. A good determination of the instrumental corrections after the close of telegraph work, is far preferable to any increase of the number of star exchanges above twenty.

When fifty stars have been satisfactorily exchanged, and on not less than three nights, the observers exchange stations, leaving the transit instrument and clock as before, meeting, however, on one night to observe for personal equation. Fifty stars are then to be exchanged again on not less than three nights, in the new position of the observers, and a new series of observations made for personal equation.

APPLICATION OF THE ELECTRIC CIRCUIT TO ASTRONOMICAL OBSERVATIONS.

It is obvious that the electric circuit, as it has been employed in the determination of longitude, may also be employed in the ordinary business of an astronomical observatory. When Professor Locke invented his electro-chronograph, by which the electric circuit was broken, and dots were printed on paper every second by the action of a clock, Mr. Walker and many others anticipated that this would immediately supersede the old method of observing transits of stars. This new method of observation has a great advantage over the old in respect of accuracy. Astronomers have hitherto measured small intervals of time by listening to the beats of a clock or chronometer, and estimating, as well as they were able, the fraction of a second when any event has occurred, such as an occultation of a star, or the transit of a star over the wires of a telescope. The ear is, however, a very imperfect organ. While the eye readily estimates a fraction of a line with the precision of a tenth, the ear seldom distinguishes smaller portions of an interval of time than a fifth. In the opinion of

Mr. Walker, the error in the mechanical part of imprint-
ing a date on the automatic clock register, and in reading
off the record, need not exceed a hundredth part of a
second. Mr. Walker estimates that a transit over one
wire, printed by the new method, is worth four wires
observed by the old method.

Another advantage of the new method of observation
arises from the increased amount of work which can be
done in a given time. Fifteen seconds is the ordinary
equatorial interval for the wires of a transit instrument.
In the new method of observing, the equatorial intervals
may be reduced from fifteen to two seconds, or even to
one and a half. In this manner the number of bisections
in a single culmination of a star may be multiplied ten
fold, making a gain of forty fold by the new or automatic
method. This is the estimated gain from the multipli-
cation of transits over wires, and the superior precision
of each. There are, however, other circumstances which
detract from this advantage, such as the time required to
prepare the apparatus for observation, to transcribe the
printed record into figures, etc.

Before, however, the full advantage of this method
could be realized, some practical difficulties remained to
be overcome, of which the most important were the two
following :

1st. It was essential to the accuracy of the new method
that the fillet of paper, or whatever might be the surface
upon which the dots were registered, should be made
to advance with perfectly uniform motion ; and

2d. It was very important that the register should not merely secure accuracy, but should also be reduced to a compact and convenient form.

The first condition, that of uniform motion of the fillet of paper, was very far from being secured with the ordinary telegraph apparatus. In this apparatus, the only arrangement for producing uniform motion consists of a fly, which, revolving with great rapidity, and meeting resistance from the air, opposes the descent of the moving weight. Now, when in telegraphing, the graver is pressed against the surface of the paper, the friction of the machinery is increased, and by this means the velocity of the paper is diminished, and the pressure may easily be increased so as to stop the motion entirely. The very operation, therefore, of registering dots on the paper, introduces an irregularity in the motion of the fillet. Now, it is indispensable to the proposed use of the apparatus, that the electric circuit should be closed during the principal part of every second; for, when the circuit is open, it is not in the power of the operator to print a dot on the paper. The result was, that when Dr. Locke's clock came to be used in connection with the Morse registering apparatus, the graver was kept pressed against the fillet of paper about nine-tenths of every second, by which means the motion of the fillet was rendered very slow and irregular; but as soon as the circuit was broken, the paper moved forward as by a sudden impulse. The first improvement consisted in reversing the action of the graver, so that when the circuit was complete, the paper was entirely free, and a dot was made by the breaking of the circuit. The paper graduated into seconds by this arrangement, exhibited dots with long intervening spaces, as shown on page 325.

It was still necessary to introduce some nicer apparatus for regulating the motion of the surface upon which the dots were to be registered. Professor Mitchell causes his registering disc to revolve with a uniform motion by connecting it with the driving apparatus of his Munich equatorial. Professor Locke prefers, for the moving power, a centrifugal clock. Professor Bond employs a machine of his own invention, called the spring governor, and described on page 324. Professor Airy, of Greenwich, employs a large conical pendulum, revolving in a circle, the diameter of which is about equal to the arc of vibration of an ordinary second's pendulum.

Mr. Saxton's arrangement of registering upon a sheet of paper wound round a cylinder, is the most convenient which has been hitherto employed.

The electric method of recording transits has been employed at the Washington observatory exclusively since December, 1849; it was introduced soon afterward at the Cambridge, Mass., observatory, and it is now used also at the Greenwich observatory.

APPLICATION OF THE ELECTRIC CIRCUIT TO ASTRONOMICAL
USES IN EUROPE.

In December, 1849, the astronomer royal of England communicated to the Royal Astronomical Society a detailed account of "the method of observing and recording transits lately introduced in America;" and he concluded his narrative by stating that the possible advantages of this method appeared so great, that he had begun to contemplate the practicability of adopting it in the Royal observatory. He proposed to record the observations upon a cylinder, perhaps revolving upon a screw axis; and suggested that great convenience would be gained if the movement of the cylinder could be made so perfectly uniform that it could be adopted as the transit clock. He, therefore, urged strongly the importance of improvements of the centrifugal or conical pendulum clock, as the only instrument yet made which is able to do heavy work with smooth motion, and with an accuracy at present so great as to make it probable that, with due modification, the greatest accuracy may be obtained.

In June, 1853, Professor Airy reported that owing to various delays his apparatus was not yet brought into use; but that he had brought it to such a state, that he was beginning to try whether the barrel moved with sufficient uniformity to be itself used as the transit clock.

The first transits recorded by the electric method at Greenwich were made on March 27, 1854, and since that

15*

time all the transits, with trifling interruptions, have been made by the same agency, except for the very slow circumpolar stars, for which they use the same wires, but by observation with eye and ear. The paper on which the punctures are to be made is folded in a wet state upon a brass cylinder covered with a single thickness of woolen cloth, and has its edges united by glue.

The punctures are produced by two systems of prickers, which have nothing in common, except that they are carried by the same traveling frame which moves slowly in the direction of the barrel axis, while the barrel revolves beneath it. One pricker is driven by a galvanic magnet, whose galvanic circuit is completed at every second of sidereal time. Professor Airy at first intended that the completion of the circuit should be effected by the same clock (regulated by a conical pendulum) which drives the barrel. He found, however, that he could not insure such a constancy in the arc of the pendulum as would make its rate sufficiently uniform to entitle it to be considered as the fundamental clock. He therefore carried wires from the pricker magnet to the transit clock, and connected them with springs whose contact is made at every second by the transit clock. A wheel of 60 teeth is fixed on the escape-wheel axis, and the teeth of this wheel in succession make momentary contacts of the galvanic springs. The position of the springs is so adjusted that the effort of the wheel-tooth upon them occurs only when one escape-tooth has passed the sloping surface of the pallet, and the other escape-tooth is dropping upon its bearing ; so that the resistance of the springs does not affect the legitimate action of the train upon the pendulum.

The other pricker is driven by a galvanic magnet, whose circuit is completed by an arbitrary touch made by an observer's finger upon a contact piece. There are three contact pieces. One is upon the eye end of the transit circle ; the other two are upon the base plate of the altazimuth, one to be used with vertical face to the right, the other with vertical face to the left. Thus altazimuth observations are referred absolutely to the same time record as transit circle observations.

It is necessary to mark upon the revolving barrel the beginnings of some minutes; and the numeration of some hours and minutes. This is done by arbitrary punctures, given by the observer's touch. In order to guide the eye through the multitude of dots upon the sheet, lines of ink are traced by means of a glass pen, which is attached to the same frame as that by which the prickers are carried.

Wires have been inserted in the wire-plates, both of the transit circle and of the altazimuth, at intervals adapted to the rapid observation by touch. The wires of the transit circle, and the vertical wires of the altazimuth are adapted to intervals of about 42″ and 48″ of arc ; the intervals of the horizontal wires of the altazimuth do not exceed 24″ of arc. These are probably the smallest intervals that have ever been used for similar observations. The old systems of wires are not disturbed nor rendered confused; so that with the transit circle, either 7 wires may be observed by ear, or 9 by touch; and with the altazimuth, either 6 by ear or 6 by touch.

Professor Airy remarks that this apparatus is now generally efficient. It is troublesome in use : consuming much time in the galvanic preparations, the preparation of the paper, and the translation of the puncture indications into figures. But among the observers who use it, there is but one opinion on its astronomical merits—that, in freedom from personal equation and in general accuracy, it is very far superior to the observations by eye and ear.

Electro-magnetic registering apparatus has also been introduced by Dr. Lamont at the Munich observatory and is described in the work, "Beschreibung der an der Münchener Sternwarte zu den Beobachtungen verwendeten neuen Instrumente und Apparate von Dr. Lamont, München, 1851."

It is not known that the American method of observing has been introduced into any other observatories of Europe; but M. Leverrier, the Director of the Imperial observatory of Paris, has recently drawn up a Report in which he recommends to construct a large meridian circle, and to economise the resources presented by dynamical electricity, in order to assure to the observations all the precision of which they are susceptible.

EXPERIMENTS MADE IN EUROPE FOR THE DETERMINA-
TION OF GEOGRAPHICAL LONGITUDE BY THE
ELECTRIC TELEGRAPH.

During the summer of 1852, the Royal observatory at
Greenwich was put in galvanic communication with the
principal telegraph offices in London, for the purpose of
determining differences of longitude with other observatories, British and continental. The first comparison
was made between the observatories of Greenwich and
Cambridge. The electric current was made to pass
through the coils of a telegraph needle at Greenwich,
and through the coils of another telegraph needle at
Cambridge, so that the completion of the circuit at Greenwich produced a movement both in the needle at Greenwich and in that at Cambridge. The order of operations
was as follows:

At 11 P. M., Greenwich mean solar time, Greenwich
commenced by giving five signals at intervals of about 2
seconds each. The turn-plates were changed, and Cambridge responded by five similar signals. These were
merely to say "All is right." Greenwich then gave
groups of signals at intervals of 10s. to 15s., in numbers
of from three to nine signals in a group (some of them
being transits of stars) to 11h. 15m. Notice was previously given of the number of signals to be expected in
each group, by giving the same number of warning signals
at intervals of about two seconds. Then Cambridge gave
similar groups of signals to 11h. 30m. Then Greenwich

gave signals to 11h. 45m., and Cambridge to 12h. 0m. This closed the night's signals. From 135 to 150 efficient signals were given. The evenings selected for the determination of the longitude of Cambridge were those of May 17 and 18, 1853. On May 17, Mr. Dunkin observed transits and galvanic signals at Greenwich, and Mr. Todd observed at Cambridge. On the morning of the 18th, the observers were interchanged, and Mr. Todd observed transits and signals at Greenwich, while Mr. Dunkin observed at Cambridge. The errors of the transit clock were determined by two methods: method (A) in which the Nautical Almanac stars were employed, but no particular care was taken for the identity of the stars at the two stations; and method (B) in which the same stars were observed at both stations, but no attention was given to the accuracy of the assumed right ascensions.

For comparison of the Cambridge transit clock, with the chronometers used at the railway stations, Professor Challis employed three chronometers. It was found necessary to reject the comparisons of one of them.

The results of these operations were as follows:

	Method (A).	Method (B).
May 17, by 146 signals, East longitude of Cambridge	22·660s.	22·601s.
May 18, by 135 signals, "	22·713s.	22·782s.
Mean	22·687s.	22·691s.
Mean of the whole		22·689s.

The result arrived at in 1829 by transmission of chronometers was 23·54s.

On May 25, 1853, signals were passed in the same manner to and from Edinburg, for the determination of the longitude of Edinburg observatory. In these comparisons the following result was obtained, that when a signal is given at Greenwich by means of a Greenwich battery, the time noted for the signal at Edinburg is later than that noted at Greenwich by $\frac{1}{17}$ of a second of time, and *vice versâ*, if the signal is given at Edinburg by means of an Edinburg battery. This difference Professor Airy ascribes to two causes; first, the time actually occupied by the transmission of the galvanic pulse, which according to the American determination, would explain less than half of the difference; secondly, the circumstance that the galvanic current when it reaches the distant needle is somewhat less vigorous than when it passes the nearer needle, and the languid movement of the distant needle catches the eye more slowly and is recorded as occurring at a later time.

In August, 1853, observations were made to determine the difference of longitude between the observatory at Berlin and that at Frankfort on the Maine. M. Encke and Dr. Brünnow observed at the former station, and Dr. Lorey at the latter. The telegraph apparatus employed was Morse's. It was agreed that Dr. Lorey should announce by signal when the experiments were about to commence, after which he was to make a series of signals during the next ten minutes, it being arranged that he was to make a new signal about the beginning of each successive minute. M. Encke observed the corresponding time at the telegraph office in Berlin, upon a mean solar chronometer, while Dr. Brünnow made similar observations upon a sidereal chronometer. The ten signals being observed, Dr. Brünnow gave a signal from Berlin to announce that a new set of experiments was about to commence, and this was followed by ten successive signals as before. The following is the result for the longitude of Frankfort west of Berlin:

SIGNALS FROM FRANKFORT.

August 12,	18m. 51·79s. Encke.	18m. 51·72s. Brünnow.
28,	51·71s.	51·85s.
	18m. 51·75s.	18m. 51.79s.

Mean 18m. 51·77s.

SIGNALS FROM BERLIN.

August 12,	18m. 51.57s. Encke.	18m. 51·92s. Brünnow.
28,	51·91s.	52·13s.
	18m. 51·74s.	18m. 52·03s.

Mean 18m. 51·89s.

or on the whole, a mean difference of longitude amounting to 18m. 51·83s. The difference of the two results can not arise from the circumstance that the transmission of the signals is not instantaneous, since during the interval of transmission, the Frankfort signals ought to have arrived too late at Berlin, in which case the difference of longitude should have turned out too great; while the Berlin signals by arriving too late at Frankfort should have indicated the difference of longitude too small. In reality, however, the Berlin signals made the west longitude *greater* than its true value. These experiments consequently indicate that with the means of observation here employed, the velocity of the electric current is insensible for the distance between Berlin and Frankfort; a distance which, by the circuitous route of the telegraph, amounts to about 300 English miles.

In November, 1853, the difference of longitude between the observatories of Greenwich and Brussels was determined by means of the electric telegraph. A galvanic telegraph needle was mounted in the observatory at Brussels in close proximity to the transit-clock, nearly as in the Greenwich observatory. The signals to be made were simple deviations of the needle, produced directly by the galvanic current through the long communicating wire.

It was arranged that the observations should be divided into two series: that in the first series an observer from Brussels (M. Bouvy) should observe both the galvanic signals and the transits for correcting the transit clock at Greenwich, while an observer from Greenwich (Mr. Dunkin) made the corresponding observations at Brussels; that this series should be continued till satisfactory observations had been obtained upon at least three evenings; that the observers should then be reversed, and the second series be observed in the same manner. The signals were to occupy one hour in each evening, from 10h. to 11h. Brussels mean solar time, each hour being divided into four quarters. The contacts of wires for completing galvanic circuit were to be made at Greenwich and with a Greenwich battery in the first and third quarters, and at Brussels with a Brussels battery in the second and fourth quarters; and between the two sets of observations at each place the poles of the battery were to be reversed.

The transit clocks were corrected by two distinct methods in the same manner as in the operations for determining the longitude of Cambridge.

The final results for the difference of longitude as determined by the two methods A and B are the following:

	Method A.	Method B.
Mean of the first series	17m. 29·256s.	17m. 29·340s.
" second series	28·538s.	28·476s.
Mean of the two series	17m. 28·897s.	17m. 28·908s.

These determinations rest upon 1104 signals. The last result 17m. 28·9s. is the best that can be given for the difference of longitude of the two observatories. It is however remarkable that the difference of the results given by the two series is 0·791s., which is probably due to personal equation of the observers.

The time employed by the galvanic current in passing between the two observatories, a distance of 270 miles, is 0·109s., which indicates a velocity of only 2500 miles per second. It is however to be remarked, that from Greenwich to London and thence to Ostend, the whole of the line is subterraneous or sub-aqueous, and it is considered probable that the observed retardation belongs almost entirely to this portion of the line.

In May and June, 1854, the difference of longitude between Greenwich and Paris observatories was determined; and the arrangements adopted were the same as had been previously tried between Greenwich and Brussels. Mr. Dunkin, assistant at Greenwich observatory went to Paris, and M. Faye, assistant at the Paris observatory, went to Greenwich, and the comparisons commenced the 27th of May. The second series of observations commenced June 12th, and were made by Mr. Dunkin at Greenwich and by M. Faye at Paris. The final results for the difference of longitude as determined by the two methods A and B are the following :

FIRST SERIES.

	Number of signals.	Method A.	Method B.
1854, May 27,	146,	9m. 20·40s.	9m. 20·38s.
" 29,	146,	20·59s.	20·56s.
" 31,	147,	20·54s.	20·56s.
June 3,	145,	20·45s.	"
" 4,	124,	20·49s.	20·53s.
	Means	9m. 20·49s.	9m. 20·51s.

SECOND SERIES.

	Number of signals.	Method A.	Method B.
June 12,	132,	9m. 20·77s.	9m. 20·76s
" 13,	132,	20·79s.	20·77s.
" 17,	140,	20·77s.	20·75s.
" 18,	137,	20·69s.	20·73s.
" 20,	148,	20·74s.	20·75s.
" 21,	155,	20·79s.	20·74s.
" 24,	151,	20·84s.	20·84s.
	Means	9m. 20·77s.	9m. 20·76s.
	Concluded longitude	9m. 20·63s.	9m. 20·63s.

In order to refer the position of the observatory of Greenwich to the ancient meridian of France, we must subtract from the preceding result 0·12s. which represents

the distance between this meridian and the present position of the transit instrument of the Paris observatory.
We thus have for the final result 9m. 20·51s.

The difference of longitude between Greenwich and
Paris has been heretofore determined with all the
accuracy which science could supply; by eclipses of
the sun; by occultations of stars; by explosions of
rockets; by geodetic triangulation; and by the transportation of chronometers.

The first important measure was made in 1790, by a
triangulation conducted by General Roy for England,
and by MM. Cassini, Mechain, and Legendre for France.
This measurement gave the difference of longitude
9m. 18·8s.

The second geodetic measurement was made in 1821,
2 and 3, by Captains Kater and Colby for England, and
by French astronomers from Calais to Paris. The result
of this measurement was 9m. 21·18s.

In 1825 the difference of longitude was determined by
fire signals. The operations were conducted by Messrs.
Herschel and Sabine for England, and MM. Bonne and
Largeteau for France. The result of this trial was
9m. 21·46s.

In 1838, Mr. Dent, of London, transported twelve of
his chronometers from Greenwich to Paris, and returned
them from Paris to Greenwich, having compared them
each time with the clocks at the two observatories. The
mean of the results furnished by these chronometers was
9m. 22·1s. in going, and 9m. 20·5s. in returning.

The difference of longitude between Greenwich and Paris for nearly thirty years had been assumed to be 9m. 21s.5; and this result is now concluded to have been too great by an entire second of time.

The time occupied in the transmission of the electric current was found to be 0s.086 at Greenwich, and 0s.079 at Paris, the distance being about 300 miles.

DETERMINATION OF THE VELOCITY OF THE ELECTRIC CURRENT.

The experiments of January 23d, 1849, between Cambridge, New York, Philadelphia, and Washington, afforded an approximate determination of the *velocity* of the electric fluid. If the fluid requires no time for its transmission, then the star signals given at either station ought to be similarly printed at all the stations; and the fraction of a second registered upon any one scale should be identically the same as upon every other. But if the fluid requires time for its transmission, these fractions will be different. Suppose the clock to be at Washington; that an arbitrary signal is made at Cambridge; and that the time required for the transmission of a signal between the two places is the thirtieth of a second. Then the clock pause will be registered at Cambridge $\frac{1}{30}$th of a second after it took place and was recorded at Washington, and the arbitrary signal pause will be recorded at Cambridge as soon as it is made, or $\frac{1}{30}$th of a second before it reaches Washington. We shall thus have the interval between the signal pause and the preceding clock pause, longer at Washington than at Cambridge, and the excess on the Washington register will measure twice the time consumed in the transmission of the signals between the two stations.

Thus, in the following figure, let the upper line represent a portion of the Washington time scale, corresponding to 15, 16, etc., seconds, and the lower line the

Washington, ___15_____16___B___17_____18____.
Cambridge, _____
              15   16 A   17   18

same for Cambridge, each division being a little later than the corresponding one for Washington. Then if an arbitrary signal is made at Cambridge between 16 and 17 seconds, and printed at A, the record on the Washington scale will be at B, and the interval from 16 to B will exceed that from 16 to A by twice the time consumed in the transmission of the signals from Cambridge to Washington.

In the observations of January, 1849, Professor Walker detected a difference in the registers of the papers at the several stations, and it indicated a velocity of the electric wave of 18,800 miles per second.*

On the 31st of October, 1849, similar experiments were repeated between Washington and Cincinnati, indicating a velocity of only 16,000 miles per second.†

Dr. B. A. Gould has discussed the same experiments, and has deduced from the observations of January 23d a

---

* Proceedings Am. Phil. Soc., vol. V., p. 76.
† Astronomical Journal, vol. I., p. 55.

velocity of 18,000 miles per second, and from the experiments of October 31st a velocity of 18,330 miles per second.*

On the 12th of November, 1849, Professor Mitchell performed a series of experiments on the line between Pittsburg and Cincinnati, the circuit being formed by a wire from Cincinnati to Pittsburg, and a second wire from Pittsburg to Cincinnati, constituting a length of 607 miles. The wave time deduced from these experiments was 0s.02128, corresponding to a velocity of 28,524 miles per second.†

The importance of an accurate determination of the velocity with which signals travel along the wires of the electric telegraph, induced the superintendent of the Coast Survey to undertake a very extensive series of experiments for this purpose. On the night of February 4th, 1850, the telegraph lines from Washington to Pittsburg, from Pittsburg to Louisville, and from Louisville to St. Louis were all united, so that signals were transmitted directly from Washington to St. Louis, and recorded on the registers of the four telegraph offices. The length of the wire constituting the telegraph line was 1049 miles, and the shortest distance between the extreme stations through the ground was 742 miles. The temperature was at zero of Fahrenheit from Pittsburg to St. Louis, and at eight degrees at Washington. The insulation was so perfect, that each station could receive

---

* Proceedings Am. Assoc. at New Haven, p. 97.
† Astronomical Journal, vol. I., p. 16.

the writing of all, without change of adjustment. A clock in Washington, prepared by Mr. Saxton, graduated the time scales on the Morse registering fillets at all the stations, and arbitrary signals were given at one station, and received at all the others. Thus Pittsburg, Cincinnati, Louisville and St. Louis, were successively, for a period of ten minutes, made the stations for arbitrary signals, which were printed on all the registers every three seconds. These observations have been carefully analyzed by Mr. S. C. Walker, by Dr. B. A. Gould, and by Mr. R. Kullman, of the Bavarian engineers. Mr. Walker's final conclusion is, that the velocity of the galvanic wave in the iron wires of the telegraph lines is 15,400 miles per second; and that the velocity of the wave in the ground is not more than two thirds of the velocity in the iron wires.* Dr. Gould obtained as the most probable result, a velocity of 14,900 miles per second through iron wire, and he concluded that the signals were in no case transmitted through the ground.† Mr. Kullman obtained a mean result of 13,000 miles per second in iron wire, and 9,740 in the ground.‡

From the experiments of February 5, 1850, between Washington and Charleston, Dr. Gould deduced the velocity of the electric current equal to 16,856 miles per second.§

* Coast Survey Report for 1850, p. 87.
† Proceedings Am. Asso. at New Haven, p. 92.
‡ Ibid., p. 401.   § Ibid., p. 97.

On the 8th of July, 1850, Mr. Walker tried a variety of experiments, for the purpose of testing, by means of the chemical telegraph, the velocity of propagation of the electric current indicated by the Morse telegraph lines. These experiments were performed on the line from Boston to New York, on a circuit of 407 miles in length, 220 of which were of iron wire in the air, and 187 were through the ground. The marks were recorded on paper, previously moistened with a solution of ferro-cyanate of potassa. As there was no astronomical clock in connection with the line, Mr. Walker tapped at intervals of two seconds on the make-circuit key, and thus graduated the chemical disc of paper at Boston and New York. The operator at the New York office imprinted in every third interval, between the marks of the graduated scale, three short marks, which were also recorded at both stations. These experiments, 63 in number, indicated a difference of the New York marks on the Boston time-scale, of about one second for every 12,000 miles.*

The experiments made in July and August, 1852, between Washington and Petersburg, Va., indicated a velocity of the electric current equal to 9,800 miles per second.†

Numerous experiments have been made in France by MM. Fiseau and Gounelle in 1850, and by MM. Bur-

---

* Astronomical Journal, vol. I., p. 108.
† Coast Survey Report for 1852, p. 26.

nouf and Guillemin in 1854, to determine the velocity

of the electric current. The following is the principle

upon which these experiments are founded. Conceive an

insulated metallic wire, a hundred miles or more in

length, the two extremities of which are brought close

together. Near one extremity of the wire, but not in

contact with it, is one of the poles of a battery, of which

the other communicates with the ground. Near the other

extremity of the wire, but not in contact with it, is the

wire of a galvanometer, of which the other end com-

municates with the earth. If at the same instant we

touch one end of the wire to the battery and the other

to the galvanometer, the current runs through the

wire, reaches the galvanometer, and deflects the needle.

But the current requires a certain time to traverse the

wire. If the contacts continue sufficiently long, the cur-

rent will reach the galvanometer, and deflect the needle.

If the contacts do not continue sufficiently long, the

current will not reach the galvanometer, and the needle

will not be deflected. By gradually diminishing the

time of contact, we may determine the exact interval at

which the deviation ceases. This interval is the time

which the current requires to traverse the wire.

The contacts are made by the rapid revolution of a
wheel whose circumference consists of narrow strips of
wood and brass alternately.

MM. Guillemin and Burnouf concluded from their
experiments that the velocity of the electric current in
an iron wire one sixth of an inch in diameter, was

108,900 miles per second.*   MM. Fiseau and Gounelle deduced a velocity of 112,680 miles per second in copper wire one tenth of an inch in diameter, and 62,600 miles in iron wire one sixth of an inch in diameter.†

All these results differ materially from that obtained in 1836 by Professor Wheatstone, who determined the velocity of frictional electricity to be 288,000 miles per second.   Professor Faraday is of opinion that the velocity of discharge through the same wire must vary with the tension or intensity of the first urging force ; and on account of the lateral induction of the current, he considers that the velocity must be different if the wires be turned round a frame in small space, or be spread through the air through a large space, or adhere to walls, or be laid upon the ground.

The telegraph wires from London to Manchester are covered with gutta-percha inclosed in metallic tubes, and buried in the earth; and when they are all connected so as to make one series, form a length of over 1500 miles. Professor Faraday placed one galvanometer at the beginning of the wire ; a second galvanometer in the middle ; and a third at the end ; the three galvanometers being side by side in the same room, and the third perfectly connected with the earth.   On bringing the pole of a battery into contact with the wire through the galvanometers, the first galvanometer was instantly affected; after about a second the next was affected; and it required two seconds for the electric stream to reach the

---

* Comptes Rendus, Aug. 14, 1854.   † Ibid., April 15, 1850.

last galvanometer.   Again, all the instruments being deflected, when the battery was cut off, the first galvanometer instantly fell to zero; but the second did not fall
until a little while after; and the third only after a still
longer interval—a current flowing on to the end of
the wire, while there was none flowing in at the beginning.

DIFFERENCES OF DECLINATION RECORDED BY ELECTRO-MAGNETISM.

Differences of declination may be recorded by means of
electro-magnetism.   This is accomplished by inserting in
the focus of the meridional telescope two systems of
spider lines, one vertical, and the other inclined at an
angle of 45°.   Let A B represent the horizontal wire of

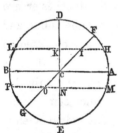

the transit instrument, D E the
middle vertical wire, and F G a
wire inclined to the latter at an
angle of 45°.   Let the telescope
be pointed upon a star as it approaches the meridian, and let it
be bisected by the wire A B, while
the time of passing the vertical wire D E is recorded.
Let the telescope remain firmly fixed in its position, and
suppose a second star enters the field at H and traverses
the path H L.   Let the instant of passing F G at I, and
D E at K be recorded.   Then if the angle D C F is 45°,
C K (which is the difference of declination of the two
stars) will be equal to K I.   The line K I is measured by

the time required for the star to describe this portion of
its path; and the observed time is easily converted into
arc of a great circle. If a third star enters the field at
M, and crosses the wire D E at N, and F G at O, then
C N is the difference of declination of the first and third
stars; and in the same manner, by observing the transits
of any number of stars over the wires D E and F G, in
the same position of the telescope, we shall obtain their
differences of declination as well as of right ascension.
In order to diminish the errors of observation, we introduce a large number of inclined wires, at intervals of two
or three seconds from each other, as well as a large number of vertical wires; and the times of transit over each
system of wires are recorded by electro-magnetism.

This method is well adapted to the construction of a
catalogue of stars, where it is proposed to record the
position of every star within the range of the telescope.
For this purpose the telescope is firmly clamped, and remains fixed in its position during the observations of an
entire evening or night, while the observer, sitting with his
eye at the telescope, has but to press his finger upon a key
at the instant a star is seen to pass each wire of the two systems already mentioned. This mode of observation has been
practiced at the Washington observatory since 1849. The
wires for right ascension are 35 in number, and are divided
into groups or fascicles of five each, the interval between
two wires being from two to three seconds. To complete
a set of observations on any one fascicle requires only
from eight to ten seconds. The wires for differences of

declination are also 35 in number, and are arranged in groups of five each. In order to prevent any con-

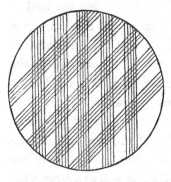

fusion between observations for right ascension and those for declination, the rule is, to observe for right ascension on one fascicle of wires first; then by a telegraphic symbol, to denote the magnitude of the star; and afterward to observe it on a fascicle of inclined wires for declination. The several fascicles are distinguished from each other by the inequalities of the intervals.

Professor O. M. Mitchell has invented a different method of registering the declinations of the heavenly bodies by means of the electric circuit, dispensing entirely with the use of a graduated circle. For this purpose, he attaches firmly to the axis of his transit instrument by a strong clamp collar, a light bar, about six feet in length. To the upper part of this arm an electro-magnet is fixed, which operates a double lever armed with a steel record-ing pen. To receive the record, a metallic plate is placed vertically on the face of the transit pier, moving in ways parallel to the circles described by the recording pen. To use this instrument for record, the observer sets for his standard star; the arm is then brought to the vertical and clamped. The instrument is then clamped, and a tangent screw gives to the observer his slow motion for bringing the star to the declination wires. When the star is bisected, the observer strikes the key with his finger, the circuit is formed, the electro-magnet brings the pen in contact with the record plate; and while in contact, the plate descends in its ways, and a zero line is described on the plate, from which all differences of declination are afterward read. Professor Mitchell uses three declination wires, and there are three zero lines obtained from the standard star. When the difference of declination of one star has been recorded, the plate moves upward about the tenth of an inch, and is ready for the next record.

To read the record of declination, the plate is laid on a carriage and leveled by four screws: a movable arc, divided into equal parts, whose values have been absolutely determined and tabulated, is adjusted so that its zero coincides with the zero line of the record. It then glides on its ways, parallel to and just above the record plate. A micrometer screw, and microscope, with a spider's web, read the fractions of the equal parts into which the arc is divided, with great facility and with great accuracy.

SECTION V.

AMONG astronomical publications in this country, the translation of La Place's Mecanique Celeste, by Bowditch, deservedly holds the first rank. Although in name merely a translation of a foreign book, with a commentary, it has many claims to the character of an original work.

The observations made by Lieutenant Gilliss at Washington from 1838 to 1842, have been published by order of Congress, and form an octavo volume of 672 pages. Three volumes of observations, made at the Naval observatory at Washington, have been published. The observations for 1845 constitute a quarto volume of 550 pages, with 13 plates; the observations for 1846 constitute a quarto of 676 pages; and the observations for 1847 constitute a volume of 480 pages, accompanied by 44 plates, showing a series of observations of solar spots by Professor Sestini, made at Georgetown observatory.

In 1852 was published No. 1 of the "Annals of the Georgetown Observatory," being a quarto volume of 216 pages, chiefly occupied with a description of the building and instruments.

In 1855 was published Vol. I., Part II., of the "Annals of Harvard College Observatory," being a quarto volume of 416 pages, containing a catalogue of 5,500 stars situated between the equator and 0° 20' north declination.

With the preceding exceptions, the American contributions to astronomical science are to be found in periodicals and the transactions of scientific societies.

The Transactions of the Royal Society of London contain some observations by American astronomers before the Revolution. The Transactions of the American Philosophical Society contain valuable papers from Rittenhouse, Ewing, Smith, Ellicott, Dunbar, Lambert, Adrain, Hassler, Gummere, Talcott, Courtenay, Loomis, Mason, Nicollet, Walker, Kendall, Bartlett, Gilliss, and several others. Among the subjects of these communications may be enumerated the transit of Venus in 1769; the transit of Mercury in 1769; the comets of 1770, 1807, 1842, 1843, and 1844; the solar eclipses of 1791, 1803, 1806, 1831, 1834, 1836, and 1838; numerous occultations of stars; moon culminations; observations of nebulæ; observations and computations for the latitude and longitude of numerous places in this country.

The Memoirs and Proceedings of the American Academy contain important papers from Willard, Williams, Winthrop, Webber, Dean, Bowditch, Fisher, Paine, W. C. Bond, G. P. Bond, and Graham. Among the subjects of these papers may be enumerated observations of the transits of Mercury in 1782, 1789, and 1845;

16*

the comets of 1807, 1811, 1819, 1845, 1846, and 1847 ; the solar eclipses of 1780, 1781, 1782, 1791, 1806, 1811, 1845, and 1846; various occultations of stars; observations of nebulæ; and observations and computations for the latitude and longitude of various places in the United States.

The Memoirs of the Connecticut Academy contain observations of the comets of 1807 and 1811, by Mansfield and Day, and the calculation of the longitude of Yale College.

The Transactions of the Albany Institute contain a notice of the solar eclipse of 1806, by Simeon De Witt, and observations of the solar eclipses of 1831 and 1832, by Professor S. Alexander.

The American Journal of Science contains some original observations of comets and eclipses, and has been the vehicle for the diffusion of much valuable information respecting subjects of passing interest.

The American Almanac, which has been published regularly since 1830, has given each year very full computations of all visible eclipses, and the elements for the calculation of occultations of stars by the moon. These computations were made by Mr. R. T. Paine until the year 1841, and since that time by Professor Pierce and Mr. G. P. Bond.

The United States Almanac, which only continued for three years, gave, in addition to the usual astronomical articles, a great variety of tables useful to computers.

The computation of the occultations of all stars down

to the sixth magnitude, for nearly twenty years, has been made by Messrs. Walker, Downes and Paine. Mr. Downes' computations for the years 1848, 1849, 1850 and 1851, have been published by the Smithsonian Institution. They contain the times of all the occultations visible at Washington, and elements for facilitating a similar computation for any part of North America.

During the session of 1849, Congress made an appropriation of $6,000 for the commencement of an American Nautical Almanac. Lieutenant (now Commander) Charles H. Davis, of the United States Navy, was appointed superintendent, and the preparation of different parts of the work was assigned to a corps of computers. Lieutenant Davis secured the valuable services of Professor Peirce as consulting astronomer; the theoretical part of the work was placed under his direction; and most of the calculations pass under his final revision.

The first volume was published in 1852, being the almanac for 1855, consisting of 552 octavo pages. The first part of the work is appropriated to nautical purposes, and is calculated for the meridian of Greenwich. The second part is designed for the promotion of astronomical science, and is adapted to the meridian of Washington. The nautical part consists of an ephemeris of the sun and moon, and of the planets Venus, Mars, Jupiter and Saturn, together with tables of lunar distances. The ephemeris of the moon is calculated from new tables founded on Plana's theory. The ephemeris of Mercury is derived from the theory given by Le Verrier; the ephemeris of Venus is founded on Lindenau's tables, with corrections from the labors of Breen, Airy and Le Verrier. The ephemeris of Uranus is calculated from Bouvard's ellipse, combined with Le Verrier's perturbations by Jupiter and Saturn, and Peirce's perturbations due to Neptune. The ephemeris of Neptune is founded on Walker's orbit and Peirce's perturbations.

The Almanac for 1856 was published in 1853. It constitutes a volume of 574 pages, and is prepared upon nearly the same plan as the preceding volume. The Almanac for 1857 was published in 1854; and that for 1858 was published in 1855.

The first periodical undertaken in this country devoted exclusively to astronomy, was the Sidereal Messenger, edited by Professor Mitchell. This was designed to exhibit in a popular form the recent discoveries in astronomy, and by this means to cultivate a more general taste for astronomical science. The work was commenced in July, 1846, and continued for a little over two years, when it was abandoned for want of patronage.

At the meeting of the American Association for the Advancement of Science at Cambridge in August, 1849, Professor J. S. Hubbard presented a paper on the establishment of an astronomical journal in the United States, and the subject was referred to a select committee. The proposition was generally approved, and the first number of the "Astronomical Journal" was issued in November, 1849, under the editorship of Dr. B. A. Gould,

This journal is devoted exclusively to the publication of original researches and observations in astronomy, geology and kindred branches. It is conducted upon the model of the *Astronomische Nachrichten* of Professor Schumacher, and the numbers appear at irregular intervals, as matter accumulates, or important information is received. A volume consists of twenty-four numbers, each containing eight octavo pages. Vol. I. was completed in April, 1851. Vol. II. was completed in September 1852. Vol. III. in June 1854; and No. 22 of Vol. IV. was published in June, 1856. The numbers have accordingly averaged a little more than one per month.

This journal has attained a high reputation, and has imparted a fresh impulse to the cause of science in the United States. It contains numerous observations of newly discovered comets and planets made in Europe as well as the United States, together with remarks respecting the orbits of these bodies and various questions in astronomy and the pure mathematics. Notice of the first discovery of a comet or a planet is immediately announced by a special circular, so that the attention of observers throughout the country is immediately directed to these bodies.

The tables from which the lunar ephemeris in the Nautical Almanac was computed, have been published in a quarto volume of 326 pages. These tables were constructed from Plana's theory, with Airy's and Longstreth's corrections; with Hansen's two inequalities of long period arising from the action of Venus, and Hansen's

values of the secular variations of the mean motion
and of the motion of the perigee ; and they are arranged
in a form designed by Professor Peirce.

Mayer's Lunar Tables, which were published in 1753,
represented the moon's place with greater accuracy than
any which had hitherto been constructed. The number
of arguments used in calculating the moon's longitude
was fourteen.   These tables received the approbation of
the British Board of Longitude, and the widow of Mayer
received on account of them a considerable sum of
money from the British government.

In 1780 were published Mason's Tables of the Moon.
In their construction and arrangement they resembled
Mayer's tables, but the number of arguments employed in
calculating the moon's longitude amounted to twenty-two.

In 1806 Bürg's Lunar Tables were published under the
auspices of the French Bureau des Longitudes.   The
number of arguments employed in the calculation of the
moon's longitude was twenty-eight.

In 1812, Burckhardt's Lunar Tables were published.
The number of arguments in the moon's longitude was
thirty-six.

In 1824, appeared Damoiseau's Tables of the Moon,
founded solely on his own theoretical researches.   The
number of arguments in the moon's longitude is forty-
seven.

The American Lunar Tables are constructed upon a
plan recommended by Carlini.   The number of arguments for the moon's longitude is seventy-nine.

SECTION VI.

THE MANUFACTURE OF TELESCOPES IN THE UNITED STATES.

VARIOUS attempts have been made in this country to manufacture both reflecting and refracting telescopes. I shall speak of each of them in succession.

REFLECTING TELESCOPES.

A great many reflecting telescopes have been constructed by amateur astronomers in different parts of the country; but, for the most part, these attempts have been but moderately successful, and have contributed but little, if any thing, to the progress of science. The most important exception to this remark was in the case of a telescope manufactured in 1838, by Messrs. Smith, Mason and Bradley, the two former gentlemen being at that time students of Yale College. This telescope had an aperture of twelve inches, and a focal length of fourteen feet. The mirror was cast, ground, and polished by their own hands. Stars of less than one second's distance, were separated by this instrument; the faint star, "debilissima," near ε Lyræ, was easily shown; and the nebula in Hercules, between η and ζ, was resolved into an immense number of small stars. With this instrument, Mr. Mason made some very accurate observations of three nebulæ, of which an account is given in the Transactions of the American Philosophical Society. This paper affords but a foretaste of what might have been anticipated from the talents of Mr. Mason, had not his course been arrested by his premature death, which occurred Dec. 26th, 1840.

Several mechanics have undertaken the manufacture of reflecting telescopes for sale, but the only one who has pursued this business to any great extent is Mr. Amasa Holcomb, of Southwick, Massachusetts. Mr. Holcomb first attempted the grinding and polishing lenses about the year 1826. He then proceeded to the manufacture of refracting telescopes, but being discouraged by the difficulty of obtaining suitable glass, he turned his attention to reflectors. In this he succeeded remarkably well, and now his telescopes are found in almost every State of the Union, and some have been ordered for foreign countries. Mr. Holcomb now manufactures four sizes of instruments.

The first size is 14 feet long and 10 inches aperture, with six eye-pieces, magnifying from 100 to 1000 times.

The second size is 10 feet long and 8 inches aperture, with six eye-pieces, magnifying from 60 to 800 times.

The third size is $7\frac{1}{2}$ feet long and six inches aperture, with five eye-pieces, magnifying from 40 to 600 times.

The fourth size is five feet long and four inches aperture, with four eye-pieces, magnifying from 40 to 300 times.

These telescopes are of the Herschelian form, and have received medals from the American Institute of New York, and the Franklin Institute of Philadelphia, after a most thorough and severe examination. With a telescope of the second size, the double stars, 51 Libræ, and ζ Bootis, the components of which are distant from each other but little more than one second, have been easily separated, and Saturn's ring seen double nearly throughout its visible portion. Mr. Holcomb has sold five telescopes of his first size, and as many of the second, with a much larger number of the smaller sizes.

REFRACTING TELESCOPES.

The experiments which have been made in this country in the manufacture of refracting telescopes, may be divided into two classes : namely those which have employed American glass, and those which have employed foreign glass.

Several telescopes of small dimensions have been made of American glass, which have performed quite satisfactorily; but the attempts to make large telescopes with American glass, so far as the results have been laid before the public, have invariably proved failures. At several establishments in this country, glass is manufactured which answers perfectly all the ordinary purposes of the arts, and for transparency, compares well with foreign glass; but it has been found impossible to obtain large discs possessing that entire homogeneity and freedom from veins which are demanded in a lens in order that it may produce a perfect image.

In the years 1846 and '48, Mr. Alvan Clark, of Boston, made two telescopes of East Cambridge flint-glass, having an aperture of five inches, which will show the division of the close pair in Zeta Cancri, and Zeta Bootis, whose distance is about one second. He has, however, expressed his determination to make no more telescopes of American glass, until he can find specimens of a better quality. It may be safely asserted, notwithstanding some pretensions to the contrary, that no *good* telescope of large dimensions has yet been manufactured of American glass.

Ever since the invention of the achromatic telescope by Dollond, about a century ago, one of the greatest obstacles to the construction of large telescopes, has been the difficulty of obtaining large discs of glass of perfectly uniform density and free from veins. The chief difficulty seems to arise from the difference in the specific gravity of the constituents of glass; some melt at a lower temperature, and sinking through the mixture, leave a streak in descending; some decompose in a heat required for the fusion of others. It has been said that the glass employed by Dollond in the manufacture of his best telescopes was all made at the same time; and the largest achromatic object glasses constructed in England, until recently, did not exceed five inches in diameter. More than half a century ago, the English Board of Longitude offered a considerable reward for bringing the art of making flint-glass for optical purposes to the requisite perfection, but it led to no important discoveries. The Academy of Sciences at Paris, offered prizes in vain for this object; and it remained for a man, not distinguished by education, nor a glass-maker by trade, M. Guinand, of Switzerland, to have the honor of arriving at the solution of the difficulty.

Guinand was born at Brenets, near Neufchatel, and was a workman in the clock and watch trade. Having been permitted to inspect an achromatic telescope, he determined to make one for himself, but could find no glass suitable for this purpose in Switzerland. He obtained some flint-glass from England, but this was not always perfectly pure. He melted it anew, but did not obtain satisfactory glass. He then erected on the river Doubs, near Brenets, an establishment in which he constructed, with his own hands, a very large furnace, and commenced the manufacture of glass, and finally succeeded in obtaining pieces large enough for telescopes. He visited Paris in 1798, and exhibited discs of from four to six inches in diameter. He afterward discovered a method of softening pieces of perfectly pure glass, for the purpose of giving them the form of a disc. In the year 1805, Guinand was invited by Reichenbach to assist him in his optical establishment which he had founded at Benedictburn, about 40 miles from Munich. Here he remained nine years, but always in a subordinate capacity. In

1814, he returned to Brenets, and established a separate manufactory, where he made telescopes, and furnished both flint and crown-glass. In 1823, he was able to produce a disc of a foot and a half in diameter. In 1824, he exhibited at the exposition of industry at Paris, a grand achromatic object-glass, which excited the admiration of the king, who solicited the son of Guinand, then present, to invite his father to take up his residence at Paris. Unfortunately, the optician was not in a condition to remove. He died in 1825, at the advanced age of nearly 80 years.

Another individual who contributed to the reputation of the establishment of Reichenbach, perhaps even more than Guinand, was the illustrious Fraunhofer. Fraunhofer was born at Straubing, in Bavaria, in 1787, and at twenty years of age (in 1807) was received into the manufactory of Reichenbach. He here exhibited the most extraordinary talents, and introduced many improvements into the manufacture of glass, as well as in the art of polishing the spherical surfaces of large object-glasses. His crowning glory was the manufacture of a telescope of nearly ten inches aperture, which was purchased for the observatory of Dorpat, in Russia.

It has been asserted that the object glass of the Dorpat telescope was made from glass cast by Guinand; but this has been positively denied by Utschneider, who states that the glass for the Dorpat telescope was cast by Fraunhofer, after Guinand left the establishment at Benedictburn; and he also states that the glass which

Guinand made was not equal in quality to that which Fraunhofer made at a later period. There can, however, be little doubt that much of the reputation of the Munich telescopes has resulted from Guinand's experiments in the manufacture of glass. The art of making this glass is kept a secret. Many particulars of this manufacture therefore can only be conjectured. Faraday found the specific gravity of Guinand's flint-glass to be about 3·616, and that its composition was silica 44·3, oxyd of lead 43·05, and potash 11·75. It is said that Guinand's original practice was to saw the blocks of glass which he obtained at one casting, into horizontal sections, supposing that every part of the same horizontal section would have the same density. A fortunate accident conducted him to a better process. While his men were one day carrying a block of this glass on a hand-barrow to a saw mill, the mass slipped from its bearers, and rolling down a declivity was broken to pieces. Guinand selected those fragments which appeared perfectly homogeneous, and softened them in circular molds in such a manner that on cooling, he obtained discs that were afterward fit fo working. To this method he adhered, and contrived a way of cleaving his glass while cooling, so that the fractures should follow the most faulty parts. When flaws occur in the large masses, they are removed by cleaving the pieces with wedges, and then softening them again in molds which give them the form of discs.

It will be remembered that glass softens so as to be readily molded into any required shape, at a temperature much below that of complete fusion; and it appears to be requisite in this second operation of forming the glass into discs, to stop short of the melting point. If the glass be completely melted, bubbles of air rise through the glass, and are found caught in the glass after it is cooled, diminishing its transparency, and perhaps causing even worse defects. Many discs are spoiled in this manner. The advantage of allowing the glass to cool before it is cast into discs, is, that it affords an opportunity to inspect the casting, and select such portions as appear less faulty. Each fragment is then put in a separate crucible or mold, having a diameter such as it is proposed to give to the disc, and softened by heat until it accomodates itself perfectly to the mold; and some discs have marks of having-been pressed down into the molds by a weight upon the top. It is then annealed by slow cooling in the manner of ordinary glass ware.

After the death of M. Guinand, his widow and one of his sons set up works in Switzerland, upon the father's principles, and were succeeded by M. Theodore Daguet (of Soleure, near Neuchatel), who sent to the London Exhibition of 1851, several discs of flint-glass, the largest being 15 inches in diameter; and a disc of crown-glass of 7 inches diameter, which were examined and found to be good. M. Daguet, by a process of his own, gives to flint-glass a degree of hardness not attained by any other manufacturer. His glass, particularly the flint, is distinguished both by its homogeneousness and its peculiar

property of resisting all decomposition by the action of
air. A council medal was awarded to him at the London
Exhibition.

The other son of Guinand was introduced by M. Lerebours, of Paris, to M. Bontemps, who had devoted much
attention to the manufacture of glass generally, and particularly of such as is required for optical purposes. He
formed an association with Bontemps, which, however,
was not of long continuance. In 1828, they succeeded
in producing good flint-glass, and discs of from 12 to 14
inches. In 1848, M. Bontemps was induced to accept
the invitation of Messrs. Chance, Brothers & Co., of
Birmingham, England, to unite with them in the attempt
to improve the quality of glass. They have succeeded
in producing a disc in flint of 29 inches in diameter,
weighing 200 pounds, and of crown-glass up to 20
inches. The former disc was exhibited at the London
Exposition of 1851, and was found to be entirely free
from any striæ, except a small portion near one of its
edges. A council medal was awarded to Messrs. Chance
for this disc.

The following is a list of prices by Chance, Brothers
& Co., of Birmingham, for warranted first quality discs
of flint or crown-glass:

4	inches	diameter,	£2 00s.	6	inches	diameter,	£7 4s.
4¼	"	"	2 7	6½	"	"	9 10
4½	"	"	2 15	7	"	"	12 00
5	"	"	3 15	7½	"	"	14 10
5½	"	"	5 00	8	"	"	17 00

8¼	inches diameter,	£19 15s.	12	inches diameter,	£44 00s.
9	"	"	22 10	12½	"	"	50 00
9½	"	"	25 10	13	"	"	57 15
10	"	"	28 10	13½	"	"	65 00
10½	"	"	31 10	14	"	"	75 00
11	"	"	35 00	14½	"	"	85 00
11¼	"	"	39 10	15	"	"	100 00

M. Maës, of Clichy, near Paris, exhibited at the London, and also at the New York Expositions, specimens of a new kind of glass, the basis of which is the oxyd of zinc, a certain quantity of boracic acid being added. Its extreme limpidity, and total freedom from color, and, so far as appears, from veins and striæ, seem eminently to fit it for optical purposes; but this glass has not stood as yet sufficient time to determine its real value. A prize medal was awarded to M. Maës at the London Exposition.

The establishment of M. Guinand, at Paris, is now conducted by M. Feil, grandson of P. L. Guinand, and the following are the prices at which he furnishes discs of either crown or flint-glass of the first quality for telescopes:

4	inches diameter,	60 francs.	10	inches diameter,	500 francs.
5	"	"	100 "	11	"	"	550 "
6	"	"	200 "	12	"	"	600 "
7	"	"	250 "	14	"	"	1000 '
8	"	"	400 "	16	"	"	2000 "
9	"	"	450 "	20	"	"	5000 "

Mr. Joseph Baden, of Kohlgrub, in Bavaria, was formerly a workman in the establishment of Utschneider, at Munich, but for many years has conducted an establishment on his own account. He makes large discs both of flint and crown-glass of the very best quality for telescopes.

While experiments, made in this country with American glass, have generally proved failures, experiments with the aid of foreign glass have been more successful. Three artists have specially distinguished themselves in the manufacture of refracting telescopes, viz., Mr. Henry Fitz, of New York; Mr. Alvan Clarke, of Boston; and Mr. Charles A. Spencer, of Canastota, New York.

TELESCOPES BY HENRY FITZ, OF NEW YORK.

Mr. Fitz's first telescope was a Cassegrain reflector of six inches aperture, and three feet focal length, which was constructed in 1838. In 1844 he saw the fine Munich telescope of the Philadelphia High School observatory, and determined to attempt the construction of an achromatic. In this he succeeded by first making a lens of three inches aperture, and afterward one of $3\frac{3}{4}$ inches, being the largest piece of flint glass he could obtain. The quality of both of these lenses was impaired by veins, as the concave lens was of quite ordinary table-ware glass; still, they compared so favorably with good Munich telescopes, that Mr. Fitz immediately commenced one of six inches aperture. This was also filled with striæ, excepting in the convex lens, which (like the first two) was made of French mirror plate. This telescope was examined by the late S. C.

17

Walker, and by Professor Kendall, at the Philadelphia High School, and elicited high approbation. In 1849, Mr. Fitz completed a telescope of $6\frac{3}{8}$ inches aperture for the use of the Chilian expedition, and this was the first telescope composed of proper crown and flint discs, all his previous telescopes having been made of French mirror plate convex lenses, instead of the superior optical crown, of which he was hitherto ignorant. Since 1849, Mr. Fitz has been continually increasing the size of his object-glasses, until he has at last attained to the dimensions of the largest instruments furnished by Merz and Mahler, of Munich. We shall enumerate the principal telescopes which have been furnished by Mr. Fitz, commencing with those of the largest size.

No. 1 has a clear aperture of $12\frac{1}{2}$ inches, and a focal length of 17 feet. It has 7 negative and 6 positive eye-pieces, the highest magnifying power being 1200. The declination circle is 20 inches in diameter, graduated to 20′, and reads by four verniers to 20″. The right ascension circle is 20 inches in diameter, graduated to 20′, and reads by two verniers to two seconds of time. The telescope is moved by clock-work, and is furnished with a micrometer. This telescope was sold to the Michigan University for $6000. Dr. Brünnow, the director of the Michigan observatory, pronounces this telescope to be a good one, and says that it compares favorably with the Munich instruments of large size. The six stars in the trapezium of Orion are visible without difficulty, and Enceladus appears well at all times. The discs of the planets, and even the brightest stars, are very well defined.

No. 2 has an aperture of 9¾ inches, and a focal length of 14 feet. It has 7 negative and 6 positive eye-pieces, the highest magnifying power being 1000. The circles are of the same size as in No. 1. This telescope was sold to West Point Academy for $5000.

No. 3 has an aperture of 9 inches, and a focal length of 9½ feet. The highest magnifying power is 600. This telescope was sold to Mr. Rutherford, of New York, for $2,200. It was made with an unusually short focus, to accommodate the size of Mr. Rutherford's dome. The performance of this telescope is highly satisfactory.

No. 4 has a focal length of 11 feet, and an aperture of 8¼ inches. It has twelve eye-pieces, the highest magnifying 800 times. Price, with clock-work and micrometer, $2,200; with plain mounting, $1,600. Mr. Fitz has sold two telescopes of this size; one to Mr. William S. Vanduzee, of Buffalo, N. Y., the other to the Friends' High School of West Haverford, Pa.

No. 5 has a focal length of 8 feet, and an aperture of 6¼ inches. Highest magnifying power 500. Price, with clock-work and micrometer, $1,300. Mr. Fitz has sold four telescopes of this size—one to Lieutenant Gilliss, for the use of the Chili expedition; a second to Mr. Vanarsdale, of Newark, N. J.; a third to South Carolina College, Columbia, S. C.; and a fourth to Dr. William F. Hickock, of Burlington, Vt.

No. 6 has a focal length of 7 feet, and an aperture of

5 inches. Highest magnifying power 400 times. Price, with clock work and micrometer, $1050; without clock-work and micrometer, $825. Several telescopes of this size have been sold.

No. 7 has a focal length of 5 feet, and an aperture of 4 inches. Highest magnifying power 250 times. Price $225, without clock-work or micrometer.

Mr. Fitz obtains his crown-glass from the manufactory of Bontemps, of Birmingham, England; his flint-glass he obtains from Paris.

Several of these instruments have been subjected to a very thorough trial before they were purchased. The instrument for the Chilian expedition was procured under the following circumstances. Mr. Fitz volunteered to make an object-glass from Guinand's discs, of the same dimensions as that of the High School observatory in Philadelphia, which should be compared with that instrument, and, if pronounced equal to it, he should charge for it only the cost of a similar lens at Munich. In May, 1849, Professor Kendall, of the High School observatory, made trial of the Fitz object-glass upon the moon, Jupiter, and several double stars; and, after careful comparison with his Fraunhofer, declared himself unable to pronounce which was the better glass. Several other competent judges assisted at the trial, and concurred with Professor Kendall in his opinion. The glass was therefore purchased by the government according to the contract. Lieutenant Gilliss, after thorough trial, pronounced this telescope perfectly satisfactory, and says that it readily shows the sixth star in the trapezium of Orion, and the daily variations in the colored portion of Mars.

TELESCOPES BY ALVAN CLARK, OF BOSTON.

A little more than ten years since, Mr. Alvan Clark, of Boston, undertook the manufacture of telescopes. His first experiments were with reflectors, but being dissatisfied with these, he attempted the manufacture of object-glasses. In 1846 and 1848, he made two object-glasses of East Cambridge flint-glass. Between them he made a telescope, of 5¼ inches aperture, of Guinand glass which was sold to Mr. Welles, of Newburyport. This telescope separated the close pair in the triple star Gamma Andromedæ, whose distance is two fifths of a second, and showed the sixth star in the trapezium of Orion at intervals, though with difficulty. After these, Mr. Clark made a telescope of 4¾ inches aperture with which he discovered three new double stars; and he has made in all more than a dozen object-glasses exceeding four inches aperture. The following is a list of the largest which he has made:

1. His largest object-glass is of eight inches aperture, and now in the hands of Rev. W. R. Dawes, of England, unsold. It was sent to England in October, 1855, and under date of December 20th, Mr. Dawes writes respecting it as follows: "Its efficiency is certainly greater than that of any other telescope I have tried, both in defining the features of a planet, and in splitting close double

stars. I have already seen Enceladus pretty steadily at the conjunctions; and on the 18th he was so plainly visible near his eastern elongation that I detected him before I had quite brought the eye-piece up to focus."

2. A telescope of 7¾ inches aperture, still on hand.

3. A telescope of 7½ inches aperture and 9½ feet focal length, was sold to Rev. W. R. Dawes and sent to England in March, 1854. The following are the remarks of Mr. Dawes respecting it:

" Though the crown-glass has a considerable number of small bubbles, the performance of the telescope is not sensibly affected by that circumstance. In other respects the materials are good; and the figure is so excellent, and so uniform throughout the whole of the area, that its power is quite equal to any thing which can be expected of the aperture; and consequently both in its illuminating and refracting powers, it is decidedly superior to my old favorite of 6⅓ inches aperture. As a specimen of its light, I may mention the companion of v Ursa Majoris as having been pretty steadily seen with it; and also that I have never seen Saturn under tolerable circumstances during the present apparition without detecting Enceladus, even when at or very near his conjunction with the planet. When exterior to or tangent to the extremity of the ring, this satellite has frequently been perceived as soon as my eye was applied to the telescope. Last spring, it was seen several times in strong twilight. In separating power, the glass is competent to divide a sixth-magnitude star composed of two equal stars, whose central distance is 0″.6."

Mr. William Lassell, of Liverpool, in a letter to the author, dated July, 1855, says of this telescope: "The optical efforts of Mr. Clark have greatly astonished me. I have had an opportunity of observing with his telescope, purchased by Mr. Dawes, and I consider it, so far as I can judge, *unsurpassed if not unequaled*."

Mr. Dawes paid $930 for this telescope, and had it fitted to his Munich equatorial stand.

4. A telescope of $7\frac{1}{4}$ inches aperture and a focal distance of 101 inches, sold to Amherst College. This telescope has a pendulum driving clock with Bond's spring governor, and is so arranged with a sector clamping upon the polar axis, that its motion is remarkably equable and firm. The circles are 12 inches in diameter; the right ascension circle reading by verniers to two seconds of time, the declination circle to 30″ of arc. The price of this telescope was $1800.

5. A telescope of $7\frac{1}{8}$ inches aperture, sold to Williams College for $900. The equatorial mounting was made by Phelps and Gurley.

6. A telescope of $6\frac{1}{4}$ inches aperture was ordered by Baron de Rottenburg, for subscribers in Kingston, Canada West. This telescope had a plain equatorial mounting and was furnished for $850.

The following is a list of the double stars which Mr.

Clark has discovered with telescopes of his own manufacture:

Right Ascension. South Declination.

	6h. 42m. 10s.	14° 58′ 47″ Mag. 6.	)	Discovered with the 4¾
8 Sextantis,	9h. 45m. 4s.	7° 24′ 1″ " 4.	}	inch object glass.
	12h. 0m. 20s.	19° 31′ 45″ " 7.	)	
95 Ceti,	3h. 10m. 42s.	1° 28′ 48″ " 5¼.	)	Discovered with the 7¼
	6h. 4m. 19s.	4° 38′ 11″ " 6¼.	}	inch, sold to Mr. Dawes.
	18h. 17m. 11s.	1° 39′ 23″ " 6¼.	)	Discovered with the Am-
	19h. 50m. 35s.	2° 38′ 1″ " 6¼.	}	herst College telescope.

Mr. Clark is now engaged on a model instrument designed to answer some of the purposes of a regular heliometer. Its micrometer will embrace two degrees, and
each spider line be supplied with an eye-piece of high
power. Its efficiency will of course depend much on the
accurate running of the driving clock, while the observer
is passing his eye from one object to the other. By re-
moving one of the eye-pieces, it becomes an ordinary
micrometer for all small distances.

TELESCOPES BY CHARLES A. SPENCER, OF CANASTOTA NEW YORK.

Mr. Spencer has long been celebrated for the excellence of his microscopes. In 1851, a committee of
the American Association for the Advancement of
Science, consisting of Professor J. W. Bailey, Dr. J.
Torrey, Professor J. Lawrence Smith, Dr. W. J. Burnett,
and Dr. Clark, made a report on Spencer's microscopes,
awarding the highest prize to his lenses, and concluded
with the remark, "the committee believe it would be an

act of injustice not to state their sincere conviction that Spencer's objectives *are now the best in the world.*"

Mr. Spencer has recently turned his attention to the manufacture of refracting telescopes, and his success in this department promises to be as great as in the manufacture of microscopes. His principal telescope is one recently completed for Hamilton College, having an aperture of 13½ inches and a focal length of nearly 16 feet. The flint and crown discs for this instrument were procured through the agents, Messrs. Cook, Beckel, & Co., New York, from Joseph Bader of Kohlgrub, and have been found to be remarkably exempt from striæ. Among the changes that have been introduced in the construction of this instrument, is a method of procuring an absolute *optical* collimation of the object-glass, combined with the usual means of making this axis coincident with the axis of the tube. The method used to produce this result enables the observer instantly to detect the existence of any error in respect to collimation; and a further advantage of the construction is that the object-glass may be made to take any angle of position to the declination axis, or to a line joining the components of a system of double stars. The focal length of the object-glass is unusually short for its aperture; and to increase its magnifying power, an equivalent of twice its focal length is obtained by the introduction of a negative achromatic near the eye-piece. The higher of the six negative eye-pieces that belong to this instrument are *solid*, and of a construction radically different from any heretofore used.

17*

They are found to be singularly free from that milkiness arising from diffused light and reflected images which invariably exist in the usual construction. The field of view appears strikingly black and impressive. Ramsden's form has been wholly discarded in the positive eye-pieces belonging to the micrometer. These have been made achromatic and orthoscopic. Each is composed of two double cemented achromatics so calculated as to give a perfectly flat field.

In the year 1855, Mr. Spencer constructed an object-glass of nine inches aperture and ten feet focal length, of discs made by Bontemps. From the few trials that have been made with it, its performance is considered excellent. Mr. Spencer has also made two object-glasses of $5\frac{1}{2}$ inches aperture and seven feet focal length, besides a large number of smaller sizes. In their construction, he has employed discs made by Bontemps, Bader, Guinand, Maës, and Daguet.

Mr. Spencer has recently received an order for a large heliometer for the Dudley observatory, at Albany, at the contract price of $14,500. The object-glass of this instrument is to be of ten inches clear aperture.

Mr. Spencer has recently visited the observatories and workshops of England, France, and Germany, preparatory to the opening of a large optical establishment at Albany, in connection with Professor A. K. Eaton and Mr. B. F. Baker.

POSTSCRIPT.

The following notice was received too late for insertion in its proper place in Chapter IV., Section I.

HAVERFORD OBSERVATORY.

This observatory is situated about nine miles west
of Philadelphia. The building is of stone, and consists
of a central part about 20 feet square, and of about the
same height, with two wings, each 15 feet square, and is
surmounted by a revolving dome 19 feet in diameter.
The instruments are an equatorial telescope; a meridian
transit circle; a prime vertical transit; a sidereal clock;
and Bond's magnetic register. The equatorial, by Henry
Fitz, has an aperture of 8¼ inches, and a focal length of
11 feet. It is mounted in the Fraunhofer style, on a
marble pedestal 8 feet high, which is supported by a
stone pier 6 feet in diameter, passing through the floors
of the building, and resting upon solid masonry 8 or 10
feet below the surface of the ground. This telescope
has an excellent spider-line, and also an annular micrometer, with five eye-pieces, magnifying from 60 to
500 times. It is provided with a clock-movement,

whose attachment is such as allows the tube to be turned while the clock is in operation.

In the west wing is placed the meridian circle, which was made by W. J. Young, of Philadelphia. It has an excellent telescope of 4 inches aperture, and 5 feet focus, with two circles 26 inches in diameter, one of which reads by four verniers to two seconds of arc; the other, is used simply as a finder. The instrument is supported by marble piers, five feet high, firmly based on masonry.

In the eastern wing is a sidereal clock by Harpur, of Philadelphia; and in the same room is the magnetic register. Upon an inward projection of the eastern wall of the center building, is mounted a prime vertical transit instrument 20 inches in length, made by Dollond. This is included in the dome containing the equatorial. The cost of the building was $2,500, and that of the instruments contained in it about $4,500. This observatory is in charge of Professor Joseph G. Harlan.

Printed in the United States
By Bookmasters